WMU Studies in Maritime Affairs

Volume 12

Series Editor

Dong-Wook Song, World Maritime University, Malmö, Sweden

WMU Studies in Maritime Affairs was launched in 2013 to inspire scholars in the field of maritime affairs to contribute to the advancement of knowledge through publications of the highest order of excellence. With this book series, the World Maritime University aims to lead an expansion of scholarly pursuits, particularly in the areas of maritime safety and environmental administration, maritime law and policy, maritime education and training, maritime energy management, ocean sustainability and governance, and shipping and port management and logistics.

Please contact the Series Editors (joma@wmu.se) for more information.

Charlott Sellberg
Editor

Simulations in Maritime Education and Training

A Companion to Simulator Instructors

Springer

Editor
Charlott Sellberg
Department of Applied Information Technology
University of Gothenburg
Gothenburg, Sweden

ISSN 2196-8772 ISSN 2196-8780 (electronic)
WMU Studies in Maritime Affairs
ISBN 978-3-032-08393-7 ISBN 978-3-032-08394-4 (eBook)
https://doi.org/10.1007/978-3-032-08394-4

This Springer imprint is published by the registered company Springer Nature Switzerland AG
The registered company address is: Gewerbestrasse 11, 6330 Cham, Switzerland

This book is dedicated to Professor Hans Rystedt, whose scholarship and collaborative spirit made it possible. In 2012, Hans was approached by a group of simulator instructors at Chalmers University of Technology. They sought his advice on how to make the best pedagogical use of simulators for maritime education. That conversation laid the foundation for over a decade of practice-based research aimed at advancing teaching and professional learning in simulation-based maritime training. Hans's intellectual curiosity and commitment to connecting educational theory with professional practice have profoundly shaped this work. His mentorship, dedication towards scientific rigor, and ability to see potential in emerging ideas have been a guiding force throughout the years. Hans, thank you for making this work possible and for your continuous support in exploring what simulation can offer to education and professional practice. But most of all, thank you for all the memorable moments we had doing research together.

Foreword: Simulations in Maritime Education and Training: A Companion to the Simulator Instructor

If we knew exactly what the future would entail, there would be little value in researching or educating ourselves to meet it. Conversely, if the future were entirely random, there would be no logic in preparing for it. Somewhere between these two extremes lies the reason why simulation-based education is essential, especially in Maritime Education and Training (MET).

Although the maritime discipline has existed for centuries, formal MET is a comparatively recent development. While substantial improvements have occurred over the past decade, the rapid pace of technological advancement, increased system integration, and growing operational complexity have introduced new challenges. These realities, along with the intrinsic nature of maritime operations, which demand a fine balance between standardization and adaptive expertise, make the need for a comprehensive, scholarly volume on simulation-based MET not only timely but urgent.

Thanks to the initiative and leadership of Dr. Charlott Sellberg, we now have such a volume. It is my distinct pleasure to contribute this preface to *Simulations in Maritime Education and Training: A Companion to the Simulator Instructor*. This book arrives at a critical moment in MET's evolution, as the field faces growing complexity and seeks to better prepare maritime professionals for the future.

Simulation in MET has often been seen as a technical solution, a means of replicating conditions, complying with regulations, and preparing students for certification. However, this perspective is incomplete. Simulation is not just about realism; it is about learning. Furthermore, learning in this context is deeply human as it involves experience, judgment, values, relationships, decision making, and reflection. That is why instructors, not machines, remain at the core of effective simulation-based training.

This volume recognizes and reinforces that truth. Drawing on years of empirical research and pedagogical reflection, the contributors highlight the obscured work of simulator instructors. They examine how simulations are designed, briefed, conducted, and debriefed. They explore how competence is assessed, not merely checked off a list, but meaningfully interpreted through professional judgment. And they advocate for instructors not just as implementers, but as co-creators of learning environments. The book highlights the urgent need to strengthen pedagogical foundations and academic integration in MET, ensuring that teaching and assessment are both professionally relevant and educationally sound.

Over the past 5 years, I have had the privilege of leading the Centre of Excellence in Maritime Simulator Training and Assessment (COAST). With insights gained from this role, I can confidently say that if there is any field where a book like this is most needed, it is MET. Simulation is already widespread in training centres around the world. What we now need is to elevate the quality of that simulation, to ensure it is not only technically sound but pedagogically rich. This volume is a decisive step in that direction.

Dr. Charlott Sellberg deserves special recognition for assembling this team of scholars and for steering the project with both academic rigor and professional relevance. Her research is significantly shaping our understanding of how simulation supports learning in professional contexts, and she continues to set new benchmarks for excellence in MET scholarship.

The chapters in this book will benefit new instructors seeking guidance, experienced instructors seeking reflection, and educational leaders aiming to raise the bar for MET. The book merges theoretical depth with practical insight in a way that few volumes manage.

To those navigating the future of maritime education, this book is for you.

Salman Nazir

Department of Maritime Operations
University of South-Eastern Norway (USN)
Notodden, Norway
Centre of Excellence in Maritime Simulator
Training and Assessment (COAST)
Notodden, Norway
Training and Assessment Research Group (TARG)
Notodden, Norway
NORD University
Bodø, Norway

Acknowledgments

I would like to express my sincere gratitude to the *International Association of Maritime Universities Simulator Group* for their time and effort in reading and providing valuable feedback on several of the chapters included in this book. Their engagement and expertise have greatly enriched the relevance and quality of this work. My heartfelt thanks also go to all the authors whose contributions have made this book possible. Your dedication, insights, and collaborative spirit have brought this volume to life. I am equally grateful to the reviewers for their constructive comments. Your willingness to share your time and expertise has been essential to the development of the book chapters.

I would also like to express my sincere appreciation to the editorial team at Springer for their professional support and guidance throughout the publication process. A special thanks goes to Professor Dong-Wook Song, Editor-in-Chief of the *WMU Studies in Maritime Affairs* book series, for his encouragement and for providing the platform to bring this work to a wider audience.

Finally, I extend our appreciation to the many maritime instructors and students who, over the years, have generously shared their training practices and reflections. Your openness and commitment to continuous learning have been a continuous source of inspiration.

In preparing the manuscripts, the Editor has made use of generative and assistive language technologies, including ChatGPT and Grammarly, for editing and proofreading purposes. All content has been reviewed and validated by the authors before publication.

Acknowledgements

About this Book

As maritime training evolves alongside technological advancements, simulator-based education plays an increasingly critical role in ensuring that seafarers meet the high standards of safety and professionalism required at sea. At first glance, one might assume that the professional expertise of experienced sea captains would serve as a guarantee for being able to conduct high-quality teaching in maritime education and training (MET). While professional knowledge is undoubtedly a prerequisite for teaching the profession and evaluating students in maritime education, academic principles such as legal certainty and equality for students are equally important for maintaining the status of maritime education as a professional qualification.

Throughout more than a decade of research and through a series of research projects involving maritime simulation, the authors of this book have examined video-recorded training and assessment situations in simulator environments, as well as engaging in interviews with maritime students, instructors, and professionals about training. Taken together, our analyses reveal the complexity of facilitating training as well as assessment of competence that not only meet the requirements of higher education examinations in a legally secure manner but also serve as the basis for issuing certification for work at sea. Overall, our research also highlights that training assessing practical competence is inherently subjective, and that it needs to be subjective, since it relies heavily on the professional judgement of experienced professionals. Therefore, additional measures are necessary to ensure that all students are given equal opportunities to participate in training and demonstrate their knowledge. There has been a growing insight throughout the years as researchers underscore the need for continuous and specified pedagogical training to better equip instructors in professional education to navigate the complex dilemmas they face in their daily work in simulated environments. At present, opportunities for professional development within higher education pedagogy remain limited for the challenges instructors face in training and assessing practical knowledge. Furthermore, there are few established guidelines in policy documents on how simulator technology should be utilized to ensure high-quality training and assessments in compliance with educational and maritime policies. One possible reason for this gap is the lack of established connections between research on learning and assessment, the practical implementation of simulator training, and the governing bodies that regulate maritime education.

These issues have also raised broader questions about the process of academic integration in MET. When maritime training programs were incorporated into the Swedish academic system in 1980, the expectation was that they would be grounded in scientific inquiry and proven experience. A 1998 evaluation by the Swedish Higher Education Authority articulated this requirement as an emphasis on increased professional knowledge, independent critical thinking, and awareness of the rapid advancements in the field through engagement with research and development activities. These aspects of professionalization are rooted in reflective analyses and research on professional practice. However, nearly three decades later, the majority of instructors in Swedish maritime education still do not hold doctoral degrees. The challenges identified in early evaluations concerning limited opportunities for academic development and subject-specific research training persist. In the Swedish context, it is notable that maritime education has not faced the same rigorous demands for scientific grounding and proven experience as programs such as nursing and teacher education. The complex challenges instructors encounter in their teaching highlight the necessity of continuous pedagogical and academic development in maritime education to sustain the processes of academic integration and professionalization. Hence, we emphasize the critical need for maritime education to prioritize the development of faculty expertise and research on maritime training to advance the field.

A potential model for such efforts can be found in Norway, where maritime universities have collaborated to establish the Centre of Excellence in Maritime Simulator Training and Assessment. Led by Professor Salman Nazir at the University of South-Eastern Norway, this center aims to advance research and education in maritime training. While the Norwegian initiative demonstrates what is possible when academic institutions prioritize faculty research and development, the global MET system has yet to establish a similar institutional framework dedicated to advancing maritime education through research-based teaching. My argument is, for MET to meet the evolving demands of the profession, there is a need for structural investments that support faculty members in pursuing doctoral studies, engaging in pedagogical training, and contributing to research on learning in MET contexts. Such efforts would not only enhance the academic standing of maritime education but also ensure that future generations of seafarers receive training that is both professionally relevant and scientifically grounded. Without these measures, there is a risk that MET will continue to fall behind other professional programs in terms of academic rigor and institutional support for research. Strengthening teaching practices and research environments within maritime education is thus a necessary step toward securing MET's long-term development and integration within the broader academic landscape.

In this context, this book is one of many steps taken in recent years to develop MET into a professionalized teaching practice. It is grounded in practice-based theories of learning to study how maritime knowledge is taught and learned in the everyday teaching practices at maritime simulator centers, as well as on interviews with experienced MET instructors. Practice-based studies provide a framework for examining how knowledge is developed, applied, and connected to professional practice through the everyday educational activities taking place in MET institutions. Rather than viewing learning as the mere acquisition of abstract knowledge, this perspective emphasizes the situated, embodied, and socially mediated nature of professional competence and expertise. In the context of MET, this means analyzing how instructors and students engage with technologies, tasks, and professional standards in the context of simulator-based learning activities. By focusing on the interactions that unfold in these instructional settings, practice-based studies offer a nuanced understanding of how maritime competence is cultivated through the continuous and deliberate work of simulator instructors.

This book builds on such an approach, using empirical analyses of instructional interactions and interviews with experienced instructors to explore how maritime competence and expertise are shaped in and through pedagogical practice. In doing so, the book contributes to a deeper understanding of the complexities of simulator-based learning and the ways in which professional knowledge is continuously negotiated and refined in training environments. Hence, by using practice-based pedagogical theories to analyze empirical data, the book aims to bridge the gap between professional learning in theory and instructional practice. We hope that the book will assist new instructors by providing the necessary tools for supporting students in becoming maritime professionals, and that it will provide some new insights into teaching and learning, even for experienced instructors. By serving as both a practical guide and an academic reflection on simulator-based teaching, the book underscores the importance of continuous learning for instructors themselves. Overall, the book advocates for a culture of pedagogical reflection and scholarly engagement within maritime training institutions, hoping to encourage that the evolving field of MET is grounded in both practical expertise and scientific inquiry.

Department of Applied IT, University of Gothenburg Charlott Sellberg
Gothenburg, Sweden

Contents

Editors and Contributors

Editors

Charlott Sellberg is an Associate Professor in IT and Learning at the Department of Applied IT, University of Gothenburg. With a multidisciplinary background in Cognitive Science and Human-Computer Interaction, she earned a PhD in Education from the University of Gothenburg in 2018. Her doctoral work examined both the theoretical and practical aspects of simulation-based training in the maritime industry. Since then, she has continued to study the role of simulations in maritime training and assessment, in areas such as basic safety training, SAR operations, and advanced ship handling, continuing to deepen the understanding of simulation as a learning environment and informing educational practices within the maritime domain.

Contributors

Daniel Erntsson Department of Mechanics and Maritime Sciences Chalmers University of Technology Göteborg Sweden

Marte Fanneløb Giskeødegård Norwegian University of Science and Technology Trondheim Norway

Susan Harrington Department of Applied Information Technology University of Gothenburg Gothenburg Sweden

Annika Lantz-Andersson Department of Education, Communication and Learning University of Gothenburg Gothenburg Sweden

Mona Lundin Department of Education, Communication and Learning University of Gothenburg Gothenburg Sweden

Charlott Sellberg Department of Applied Information Technology University of Gothenburg Gothenburg Sweden

Amit Sharma Estonian Maritime Academy Tallinn University of Technology Tallinn Estonia

Ewa Skantz Department of Education, Communication and Learning University of Gothenburg Gothenburg Sweden

Aud M. Wahl Norwegian University of Science and Technology Trondheim Norway

Pia Williams Department of Education, Communication and Learning University of Gothenburg Gothenburg Sweden

The Use of Simulators and Simulations in MET

Aud Marit Wahl and Marte Fanneløb Giskeødegård

Abstract This chapter examines the evolving role of simulators in MET, highlighting their transition from technical training tools to comprehensive learning environments. Historically grounded in apprenticeship, MET now integrates simulation to bridge theory and practice, driven by regulatory frameworks like the STCW Convention. The chapter explores various simulator modalities and how their use reflects discipline-specific pedagogies. Through a discussion of the concept of simulator fidelity encompassing technical, psychological, and interactional aspects, the chapter examines key dimensions to consider when designing simulator training scenarios to achieve the desired goals of the exercise. With reference to an identified gap in the MET literature, specifically the lack of theoretically informed learning perspectives, the chapter utilizes the fidelity discussion to motivate another type of reflection among instructors regarding the relationship between the choice of learning goals, the choice of technology, and scenario design.

Keywords Simulation-based training · Fidelity · Socio-materiality · Maritime Education and Training (MET)

1 The Role of Maritime Simulators in MET

Educating competent maritime personnel, acknowledging that this entails a mix of theoretical and practical knowledge has been a concern ever since one started formalizing nautical education. Historically, nautical education has been rooted in apprenticeship models, where seafarers gained expertise through prolonged onboard experience. Educational programs have relied on a substantial amount of practical, on-the-job vocational training to support the students' transitions from novices to

A. M. Wahl (✉) · M. F. Giskeødegård
Norwegian University of Science and Technology, Trondheim, Norway
e-mail: aud.wahl@ntnu.no; marte.giskeodegard@ntnu.no

C. Sellberg (ed.), *Simulations in Maritime Education and Training*, WMU Studies in Maritime Affairs 12, https://doi.org/10.1007/978-3-032-08394-4_1

competent professionals (Manuel, 2017; Türkistanli, 2024). In England, efforts to formalize education for seafarers began around 1850, prompted by a series of accidents that highlighted the need for structured theoretical and practical instruction. Over the subsequent 150 years, Kennerley (1993, 2000), describes a cyclical focus on education and training, reflecting the multifaceted skillset required of seafarers. Proficient navigation demands not only operational competence but also knowledge of for example, arithmetic, astronomy, and the use of navigational tools such as charts and instruments.

With increased focus on formal education, simulators emerged as a critical tool to bridge theoretical knowledge and practical skills, offering safe environments for training in high-risk scenarios (Manuel, 2017). The earliest references to simulator use in MET date back to the 1960s, focusing primarily on technical training, such as radar operation (Quick, 1961; Weekes, 1962). These early applications were largely instrumental, accentuating the acquisition of specific operational competencies. The integration of simulators into maritime education and training (MET) has evolved significantly over the past six decades, reflecting broader shifts in pedagogical approaches, technological advancements, and regulatory frameworks. The emphasis on the use of simulators in practical training also pertains to how STCW, a fundamental regulatory document in MET, specifies the use of simulators as a requirement for skill training. In some cases, the STCW code even allows seafarers to renew their certificates with simulators. Given that IMO regulations and STCW require full mission simulators for education, all educational institutions need to have access to such equipment. However, the type and amount of equipment vary amongst institutions.

Today, the use of simulators in MET plays a crucial role in shaping how students learn and develop the skills necessary for maritime professions. Kim et al. (2021) distinguish between four main categories of simulators: desktop-based simulators, full-mission simulators, virtual reality (VR) simulators, and cloud-based simulators. By identifying a 13-factor framework for evaluation, Tusher et al. (2024) show that the instructors and decision-makers put much emphasis on regulatory standards when choosing modality. Moreover, the factors were given varying weights depending on modality. VR is valued in respect to immersive training, and cloud-based versions are considered cost-effective and user-friendly. The respondents do, however, raise concerns about the cloud and desktop-based versions' pedagogical value. While the study provides valuable insight into the stakeholders' evaluation of the different factors with respect to modality, it does not put this in relation to use. Recent studies indicate that certain modalities are used more frequently than others, reflecting the signature pedagogies of each discipline. These choices, both in terms of what modalities are used and how, are often shaped by the specific needs and practices of the field. The results of a survey, where instructors at Scandinavian higher education institutions for nautical bachelor education self-reported on how often they used ten different listed forms of simulation, show that 60% use advanced simulators (e.g., full-scale bridge environments) weekly or more often. Including those using it at least monthly, this summed up to 72% of the instructors. Looking at the other modalities used monthly or more often by the instructors, we see that computers (49%), laboratories (47%), and objects/ instruments for skill training (44%) are also frequently used. The least used modalities were AR and VR, with 95% and 93% of instructors, respectively, reporting that they either never used them or did not have access. Among the remaining instructors, none reported using these technologies more than once per semester (Giskeødegård, Berg, et al., 2024).

The use of simulators for practical training can be said to be a pillar in the signature pedagogy of MET (Giskeødegård & Wahl, 2025). Signature pedagogies are discipline-specific teaching practices that reflect the specialized knowledge and skills required in a specific trade (McLain, 2022; Shulman, 2005). All professions or disciplines have their own ways of teaching and learning that are considered fundamental and effective. These pedagogies consider the real-world in which professionals work and are often context specific as they are tailored to the unique demands and challenges of a specific trade. Signature pedagogies "*implicitly define what counts as knowledge in a field and how things become known*" (Shulman, 2005, p. 4). This implies that signature pedagogies serve as a means of socializing trainees into a community of practitioners (Gherardi, 2017) as the trainees not only acquire knowledge but also develop the identity, values, and norms associated with their future or current profession. That preferences and practices differ across professions is illustrated in Giskeødegård et al. (2024) where the maritime instructors' use of simulators was compared to the practices of instructors/ facilitators within nursing and biomedical laboratory science (BLS) simulation training. The results show disciplinary variations in both frequency and choice of modality. Illustratively, advanced simulators are much more frequently used by nautical instructors (although BLS students use laboratories for practical training as often as nautical students use simulators).

2 MET Research

Türkistanli (2024) suggests that the strong emphasis on simulators in the regulative documents has influenced their important role also in the research on MET education. In a review and bibliometric analysis of the maritime research, Türkistanli (2024) finds that many of the articles focus on simulations and/or on teaching methods, and 22 of the reviewed 133 articles focus on bridge simulators. In exploring methods used, the study shows a preference for experimental designs, testing a method or tool and evaluating their effect. Human factors are identified as a dominant narrative. While the work spots a growing interest in the MET literature on learning tools and models, it is said to be lower than other educational fields.

Another study that has contributed to our understanding of the dominant trends in the topics of research on maritime simulator training is Wiig et al. (2023), who conducted a review combining quantitative topic modeling with qualitative content analysis of articles published in the period from 1961 to 2021 in three leading maritime journals (n = 87). Their review identified four main topics in the literature: (1) management of resources emphasizing team training and technical/non-technical skills; (2) academic performance which includes technical prerequisites, practice-connections necessary of learning as well as how to measure it; (3) simulator training focusing on how training is organized; (4) authentic assessment highlighting development and validation of assessment methods. Wiig et al. (2023) also highlight the dominant role of the human factor perspective.

Both Wiig et al. (2023) and Türkistanli (2024) note that while a learning perspective on MET has increased in recent years, the approaches to learning in the MET literature overall have traditionally had an instrumental approach, often focused on experiments to measure effect and on learning objectives. Wiig et al. (2023) point out that much of the literature is based on lay theories and implicit assumptions on how people learn, rather than going into dialogue with pedagogical theory. This observation also speaks to the practices of maritime instructors, per se. In Giskeødegård and Wahl (2025), simulator instructors in maritime education were asked to evaluate four different forms of resources they made use of when planning a simulator exercise on a scale from one to five (one being to a very small extent). Their report shows that relevant experience/events from work/practice (M = 4.36) and previous exercises (M = 4.42), scored much higher than pedagogical research on the use of simulation in education (M = 2.62) or research on best practice (M = 2.53). These findings underscore the vocational emphasis on practical skill training in MET, where much emphasis is put on "real life" practical experience, which has been a pillar ever since the tradition of the apprenticeship models of learning the craft.

3 An Example: The Development of Norwegian DP Training

The vignette in the embedded box below aims to give a real-life glimpse into the historical development of simulation-based training through the example of simulators for dynamic positioning in Norway. The case is built on two informal qualitative interviews, with two central stakeholders in establishing DP training in Norway—one in nautical education and the other from one of the DP technology providers and commercial training centers. This story, in addition to giving a historical glimpse into an important process of establishing DP training and certification, is a helpful backdrop for the coming discussion on factors playing into building a simulator exercise that is perceived by involved actors as providing relevant training.

The Development of Dynamic Positioning Training in Norway
The need for specialized training in dynamic positioning (DP) systems has grown significantly over the years. Initially, DP operators (DPOs) learned through manuals, onboard experimentation, and guidance from experienced officers. The first DP system was delivered to an American ship in 1961, and Norway's first DP-equipped vessel, Berg Arctic Surveyor, followed in 1971. Kongsberg Maritime began producing DP systems in 1976 and remains a major global provider.

(continued)

(continued)

As DP technology advanced, so did the need for formalized training. In the 1980s, Norway had limited access to advanced simulators. The first high-fidelity simulator, developed in Trondheim by the Ship Maneuvering Simulator (SMS), was analog and limited to night navigation. Over time, simulators evolved to support daylight navigation and specialized DP training. A key driver for this development was the use of shuttle tankers to transport oil from the Norwegian Continental Shelf, which required tailored training (Lindøe, 2009).

Initially, few training centers offered DP courses. For example, students in Ålesund had to travel to Haugesund for hands-on training. Ålesund instructors, with support from local shipowners, with substantial effort eventually established their own simulator park, incorporating systems from multiple providers. Early simulators were standalone trainers used to teach what buttons to push and the basics of how the system operated. While this was important, it limited the instructor's ability to operational training involving contextual understanding. The instructors thus emphasized telling stories from their own work experience to provide context and highlight relevance for the students.

Eventually more advanced training required simulators connected to instructor stations, enabling scenario-based learning and the introduction of unforeseen events. The possibility to simulate operational contexts and introduce unexpected challenges was seen as a major advantage. Discovering and responding to mistakes, rather than being told about them, was essential for a learning process that permits operational understanding of the work context. The complexity of these operational and task solving layers of the training advanced along with the advancement in complexity by adding visual simulations. In Ålesund they even developed custom visualization tools in collaboration with a startup for their advanced DP simulators.

Initially, DP certification was issued by technology providers as part of the contracts when customers bought DP systems. As the need to formalize the training requirements increased, the Norwegian Maritime Authority (NMA) issued Norwegian DP certificates for a while (until 2004/2005) before the Nautical Institute (NI) took over the certification in line with their certification scheme adopted in 1983 (NI, 2025). After the inclusion of DP training in the STCW in 2010 DNV also developed a DP certification process. Today both DNV and Nautical Institute offer standards, recommendations and accredits DP training programs and ensures that the simulators, course curriculums and instructors meet industry standards (DNV AS, 2014; DNV AS, 2018; NI, 2025) in accordance with guidelines set by the International Maritime Organization (IMO) and the International Marine Contractors Association (IMCA, 2024; IMO, 2011). This development has led to a substantial increase in the number of regulatory documents and requirements related to DP training.

(continued)

(continued)

New business areas emerge as DP technology is found in different types of ships for example, diving vessel, drilling rigs, cruise ships, super yachts and wind farm service vessels etc. The instructors now experience that training must be adapted to the type of operation and ship. This requires a combination of technical training, procedures, team and leadership skills. They now offer more off the shelf and tailor-made courses to their customers. For example, courses that combines DPO training in accordance with the DP schemes while also emphasis joint training of deck and engine room officers to practice team interactions and communication.

(Sources: Information from informal interviews with two key stakeholders in Norwegian MET)

4 Simulator Fidelity

As described in the vignette above, access to simulated environments enables professionals to train not only for routine operations but also for unexpected and complex situations (Wahl et al., 2020). The example also illustrates that a joint effort of both technology developers and educators is needed to design simulations that closely resemble real work settings, thereby creating authentic and meaningful training experiences. The question of what constitutes sufficient resemblance has therefore been widely debated in academic literature.

A central concept in this debate is simulator fidelity, the extent to which a simulation replicates real-world conditions (Allen et al., 1986; Dahlstrom et al., 2009; de Oliveira et al., 2022; Hamstra et al., 2014; Hontvedt & Arnseth, 2013; Hontvedt & Øvergård, 2020; Liu et al., 2009; Salas et al., 1998; Wahl & Kongsvik, 2018). Fidelity is often described as high or low, depending on the perceived realism and complexity of the simulation (Hontvedt & Arnseth, 2013; Liu et al., 2009).

There are multiple definitions of fidelity, each offering a distinct perspective on what makes a simulation effective. These concepts serve as analytical lenses, drawing attention to different dimensions that contribute to the relevance and quality of a training exercise. For clarity, the discussion below follows the categorization proposed by Hontvedt and Øvergård (2020), which identifies three key dimensions in simulator-based training design: technical fidelity, psychological fidelity, and interactional fidelity. The chapter is summarized in a textbox that highlights key considerations for simulator instructors regarding different dimensions of simulator fidelity.

5 Technical Fidelity

Technical fidelity is defined in various ways; however, most definitions highlight the physical characteristics of technology (Liu et al., 2009). In technical terms, fidelity increases as the simulated environment more closely resembles the actual physical work environment, such as replicating the layout of a ship bridge or the physical forces acting on a vessel (Hontvedt & Arnseth, 2013; Liu et al., 2009). There are both physical and functional aspects of technical fidelity, and studies suggest that both physical that both physical and functional aspects of computer technology can influence the perceived quality of simulator-based training (Allen et al., 1986; Dahlstrom et al., 2009; Liu et al., 2009). The functional aspects relate to whether the hardware and software equipment resemble real life (equipment fidelity), if it provides realistic motion cues during operations (motion fidelity) and if it provides the participants with realistic visual-audio stimuli (visual-audio fidelity or photorealism). Functional fidelity concerns the level of psychological-cognitive replication of actual devices and stimuli in the work environment, rather than just physical entities.

Technical fidelity is vital in many training designs as it supports training activities and provides necessary technological backing. Typically, this type of training focuses on technical skills required for tasks such as maneuvering a vessel, handling bridge equipment, or operating vessel systems. The ability to handle technological tools and respond swiftly to changing environments are crucial skills among mariners that must be developed through effective training. Skill acquisition is typically achieved through repeated practice of a specific task, and simulators have shown to be excellent tools for such training as illustrated in the DP example above.

There is an ongoing debate in the field of simulator training regarding the level of technical fidelity required and the extent to which the technical characteristics of a training device need to replicate an actual work environment to ensure effective learning (Hamstra et al., 2014; Hontvedt & Arnseth, 2013; Rystedt & Sjöblom, 2012; Dahlstrom et al., 2009; Liu et al., 2009; Salas et al., 1998). Salas et al. (1998) argued that there is a tendency in aviation to prioritize simulator design over human-centered training systems, based on the belief that increased simulation realism automatically leads to improved learning. According to Dahlstrom et al. (2009), a direct causal relationship between simulator fidelity and the quality and transferability of training to an actual work environment is often assumed, encapsulated in the notion that "if it looks real, it will provide good training" (Dahlstrom et al., 2009, p. 308).

These discussions highlight the need for careful consideration of technical fidelity in training design. Rather than assuming that higher realism automatically leads to better learning outcomes, instructors and designers must ask: *What level of technical realism is necessary to meet the specific learning objectives? What is the minimum fidelity required to support effective skill acquisition without overcomplicating the training environment?* And crucially, *could an overemphasis on technical detail divert attention from core competencies or overwhelm learners?* Addressing these questions is essential to ensure that simulator-based training remains both pedagogically sound and practically relevant.

6 Psychological Fidelity

The technological qualities of the equipment in the simulators—and how the participants are set to use it—are shown in the emphasis on psychological-cognitive replication of actual devices and stimuli in the work environment when discussing technical functional fidelity. It directs our attention to other forms of fidelity, psychological fidelity, and at the same time highlights the interrelated character of the various forms of fidelity. This concept has its roots in cognitive psychology, which focuses on human capabilities and internal information processing. It usually addresses human capabilities such as problem-solving, decision-making, and situational awareness (Flin et al., 2008; Salas et al., 1998).

Hontvedt and Øvergård (2020) define psychological fidelity as the design of training environments that support the development of mental models, problem-solving strategies, and sense-making. For example, in dynamic positioning (DP) training, instructors may use advanced simulators to introduce faults or unexpected events that operators must detect and respond to—scenarios that would be unacceptable to replicate on real vessels. This ability to "turn up the heat" in a safe environment is a key advantage of simulation-based training.

An essential component of psychological fidelity is the creation of a psychologically safe learning environment. Trainees must feel comfortable voicing ideas, asking questions, and admitting mistakes without fear of negative consequences. Concerns about interpersonal risks, such as fear of failure or appearing incompetent, can inhibit learning behaviors (Schein, 1996). As Edmondson and Lei (2014, p. 36) note, "people are more likely to offer ideas, admit mistakes, ask for help, or provide feedback if they believe it is safe to do so." This sense of psychological safety is critical for fostering open communication and engagement during training.

Psychological fidelity, like technical fidelity, raises important questions for training design. Instructors must consider: *How essential is psychological realism for achieving the learning objectives? Should the training scenario be adapted to individual abilities? And how can a psychologically demanding environment still ensure safety and support for all participants?* These questions are not isolated from those raised by technical fidelity. Rather, they underscore the need for a holistic approach to simulator-based training design—one that balances realism with pedagogical effectiveness.

7 Interactional Fidelity

Doing a job well requires more than technical know-how; it demands the ability to apply skills effectively in complex, real-world situations. The concept of interactional fidelity (Hontvedt & Øvergård, 2020) captures this by emphasizing the interplay between social and material aspects of work. Related ideas such as sociological fidelity (Thomas & Reeves, 2015) and social fidelity (Wahl, 2020) highlight the importance of interpersonal dynamics in simulation-based training.

While psychological fidelity highlights the cognitive aspects of the skillset often referred to as non-technical skills (Flin et al., 2008), *interactional fidelity* refers to the interpersonal dimensions. Some of the skills emphasized in this respect are communication, coordination, collaboration, adaptability, conflict resolution, and management. Those skills are an essential part of executing daily work and contribute to safe and efficient work execution in professions that are characterized by complex and risk-exposed tasks like the maritime sector (Wahl & Kongsvik, 2018).

Recent studies (Giskeødegård et al., 2024; Wahl et al., 2025) show that students often enter training with limited understanding of their future professional roles. Simulation-based learning helps bridge this gap by offering realistic, collaborative scenarios where students practice not only task execution but also teamwork and communication. Role-play, a key method in maritime simulator training, assigns participants specific roles in simulated environments, encouraging them to act and respond as they would in real-life situations. Its effectiveness depends on active engagement, which is shaped by social dynamics and the collective culture of the training environment.

Interaction fidelity puts emphasis on how learning opportunities in a simulator setting are closely related to social constructions found in a real work context (Hontvedt & Arnseth, 2013). Moreover, authentic simulations also depend on the authenticity of collaborative activities among participants (Rystedt & Sjöblom, 2012). Thus, social factors need to be incorporated into simulation design to create a realistic social context for learning (Thomas & Reeves, 2015). Rich social fidelity provides the trainees with opportunities to visualize and understand deviations from standard practices through peer and instructor feedback. The learning process can be enhanced using storytelling to share experiences and learn from each other. When emphasizing social fidelity, the trainer also acknowledges the role of social rank and how it influences interactions during the simulator session or throughout a course (Wahl, 2020).

Given these considerations, interactional fidelity raises several important questions for training design: *How important are realistic interactions to achieve the learning goals of the exercise? Do trainees have sufficient contextual understanding to engage in realistic roleplay? How can simulations be structured to support effective communication and collaboration while maintaining authenticity?* Addressing these questions is essential to ensure that simulator-based training not only reflects the complexity of real-world teamwork but also fosters meaningful learning experiences.

Key Considerations for Instructors

Technical Fidelity

Focus: Replicate the physical and functional aspects of the ship, for example:

- The simulator mirrors ship instrumentation, equipment, software, or hardware.
- The simulator replicates hydrodynamic forces and vessel movements.

(continued)

(continued)

- The simulator imitates geographic locations, other ships, or harbours.**Instructor reflections**
- How important is technical realism in achieving the learning goals of the exercise?
- What is the minimum requirement of technical resemblance to learn a task?
- How can technical realism interfere with the learning objective?**Psychological Fidelity**

Focus: Create environments that stimulate cognitive and emotional responses to enhance learning and retention, for example:

- The trainee is placed in stressful situations with a high workload.
- The trainee must make high-risk decisions under pressure.
- The trainee develops technical knowledge into situation-specific skills.**Instructor reflections**
- How important is psychological realism in achieving the learning goals of the exercise?
- Do you need to adjust the exercise to individual trainee abilities?
- How can you ensure a safe learning environment for all trainees?**Interactional Fidelity**

Focus: Create socio-technical environments that mimic team interactions and communication patterns, for example:

- The crew roles mimic the maritime social hierarchies and ranks.
- The crew uses operation-specific procedures and checklists to coordinate work.
- The crew must communicate in multiple channels in a correct manner.**Instructor reflections**
- How important are realistic interactions in achieving the learning goals of the exercise?
- Do the trainees know the work context sufficiently to carry out a realistic roleplay?
- How can you ensure realistic tools for communication and collaboration?

8 The Evolving Role of Socio-Material Dimensions

The discussion above regarding simulator fidelity demonstrates how the interest and relevance of the socio-material dimensions in simulation-based training has grown significantly over time. Initially, maritime simulators were primarily used to develop technical skills. However, following several serious maritime accidents in the 1990s—where human factors were identified as key contributors —there was a shift toward incorporating leadership and teamwork training. This shift was inspired by

Crew Resource Management (CRM) programs developed in aviation during the 1980s and introduced to the maritime domain in the 1990s (Grech et al., 2008; Praetorius et al., 2020).

A major milestone in this development was the 2010 Manila Amendments to the International Convention on Standards of Training, Certification and Watchkeeping for Seafarers (STCW), (IMO, 2011). These amendments introduced new certification requirements, including mandatory training in leadership and teamwork, as well as updated guidance for personnel operating Dynamic Positioning (DP) systems. The International Maritime Organization (IMO) responded with model courses to support these changes. IMO Model Course 1.22, Ship Simulator and Bridge Teamwork (SSBT), introduced in 2002, emphasizes both theoretical and practical components, recommending that half of the course (21 out of 42 h) be conducted in simulators to provide hands-on experience in navigation, teamwork, and crisis management (IMO, 2023). IMO Model Course 1.39, Leadership and Teamwork, followed in 2012 (IMO, 2014). Training programs aligned with these courses are commonly referred to as Bridge Resource Management (BRM), Engine Room Resource Management (ERM), or Human Element Leadership and Management (HELM).

These programs typically focus on four core non-technical skill areas: leadership, cooperation, situation awareness, and decision-making (Flin et al., 2008). While early CRM training was largely classroom-based, it has since evolved into more immersive, simulator-based formats. The work of Rhona Flin and colleagues has been instrumental in shaping CRM training across domains (Flin et al., 2003, 2008, 2016). They advocate for profession-specific taxonomies that emphasize observable behaviors critical to safe and effective performance, while excluding harder-to-measure factors such as health and stress coping (Flin et al., 2003, 2016). These taxonomies are typically structured in three levels: four overarching skill categories, each broken down into elements, with each element illustrated by behavioral markers to support assessment during simulator sessions (see Fig. 1 for an example). This structure supports both the development and assessment of non-technical skills by providing clear, observable indicators of performance.

Fjeld et al. (2018) conducted a literature review identifying five prevalent non-technical skills used by bridge officers during navigation: situation awareness, decision-making, workload management, communication, and leadership. They highlighted a knowledge gap in understanding how these skills interact and noted a lack of research treating them as part of a unified taxonomy. Their findings also revealed a research bias toward cognitive skills, with less attention given to interpersonal competencies. This is echoed by Wahl and Kongsvik (2018), who observed that much of the literature emphasizes individualistic learning theories, with limited focus on learning as a social process.

Grech et al. (2008) observed that while Crew Resource Management (CRM) training has enhanced awareness of human capabilities and limitations, early implementations were narrowly focused on front-line personnel and specific concerns such as fatigue and communication. These programs often adopted an individualistic lens, overlooking broader systemic factors. However, more recent developments

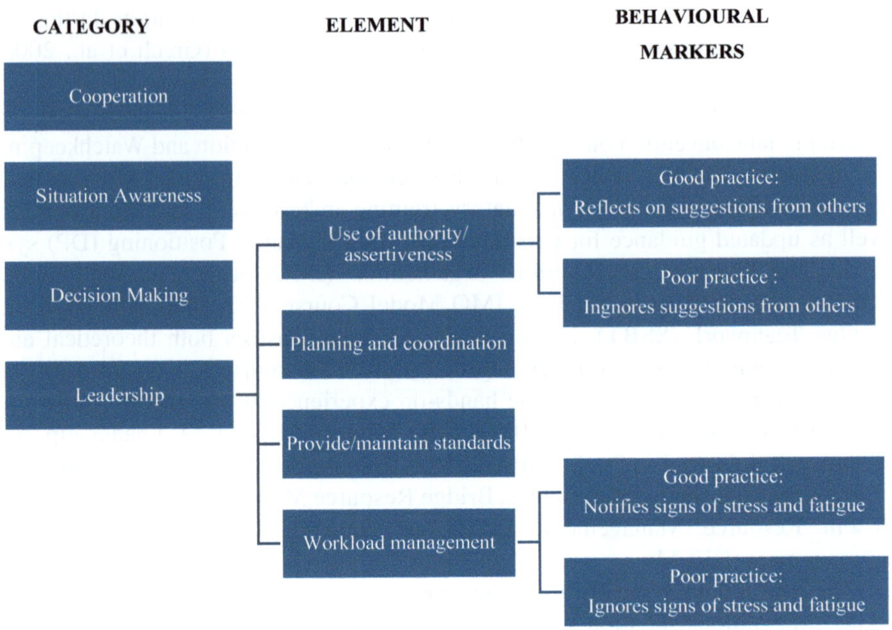

Fig. 1 Non-technical skills in maritime training based on Flin et al. (2003, 2016)

reflect a shift toward systems thinking in high-risk industries (e.g., Rasmussen, 1997; Reason, 1997), emphasizing the complex interplay between organizational levels, human behavior, and technology. This broader perspective has led to an expansion of training initiatives to include not only ship crews but also managerial and shore-based personnel, fostering a more integrated and holistic approach to safety and performance in maritime operations.

9 The Future

Technological developments that enable more distributed interaction, as well as COVID, has triggered both focus on and research about other forms of simulator exercises than the full bridge simulator (Türkistanli, 2024). One example is given by Hiroaki et al. (2022) studying the use of remote ship handling simulators for cadet training on BRM, focusing on the cadets' own evaluation of whether this is a good option for BRM training. The study captures important insights into a positive attitude amongst the cadets on such methods, which lays the ground for other, more pedagogically oriented studies to explore what this shift in modality means for the learning process. There is an increased interest for e-learning, VR, and desk-top or cloud-based solutions in MET (Kim et al., 2021; Türkistanli, 2024; Tusher et al., 2024), yet the full-scale bridge simulator is the dominant choice of simulator modality (Giskeødegård, Berg, et al., 2024; Tusher et al., 2024). At least in Norway, instructors' self-reports indicate that AR and VR-based simulators are not frequently used (Giskeødegård, Wahl, & Sellberg, 2024). As Tusher et al. (2024) show, the stakeholders evaluate the simulator modalities to have different benefits and drawbacks. Technological development laid the ground for the development of a different approach to simulation training in MET education, by stimulating a more active reflection on the choice of modality when designing exercises. The literature presented in this chapter shows that to be able to successfully do so, the instructors will benefit from actively engaging with the theoretical discussions on different forms of fidelity as well as the growing body of pedagogically oriented studies in the MET literature. Thus, this calls from the research literature on MET to also open up for new perspectives, particularly by moving beyond instrumental conceptions of training and engaging more deeply with theories of learning and professional development.

References

Allen, J. A., Hays, R. T., & Buffardi, L. C. (1986). Maintenance training simulator fidelity and individual differences in transfer of training. *Human Factors, 28*(5), 497–509.

Dahlstrom, N., Dekker, S., van Winsen, R., & Nyce, J. (2009). Fidelity and validity of simulator training. *Theoretical Issues in Ergonomics Science, 10*(4), 305–314.

de Oliveira, R. P., Carim Junior, G., Pereira, B., Hunter, D., Drummond, J., & Andre, M. (2022). Systematic literature review on the fidelity of maritime simulator training. *Education Sciences, 12*(11), 817.

DNV AS. (2014). *Certification scheme for dynamic positioning operators*. DNV -RP-0007.

DNV AS. (2018). *Competence of dynamic positioning operators*. DNV-ST-0023.

Edmondson, A. C., & Lei, Z. (2014). Psychological safety: The history, renaissance, and future of an interpersonal construct. *Annual Review of Organizational Psychology and Organizational Behavior, 1*, 23–43.

Fjeld, G. P., Tvedt, S. D., & Oltedal, H. (2018). Bridge officers' non-technical skills: A literature review. *WMU Journal of Maritime Affairs, 17*(4), 475–495.

Flin, R., Martin, L., Goeters, K., Hörmann, H., Amalberti, R., Valot, C., & Nijhuis, H. (2003). Development of the NOTECHS (non-technical skills) system for assessing pilots' CRM skills. *Human Factor and Aerospace Safety, 3*(2), 95–117.

Flin, R., O'Connor, P., & Chrichton, M. (2008). *Safety at the sharp end: A guide to non-technical skills*. Ashgate.

Flin, R., Youngson, G. G., & Yule, S. (2016). *Enhancing surgical performance: A primer in non-technical skills*. CRC Press.

Gherardi, S. (2017). A practice-based approach to safety as an emergent competence. In C. Bieder, C. Gilbert, B. Journé, & H. Laroch (Eds.), *Beyond safety training: Embedding safety in professional skills* (pp. 11–22). Springer Open.

Giskeødegård, M., Berg, H., Tveten, A. K., & Solberg, M. (2024). How instructors think about technology in simulation-based training: A survey of three programs for professional education. *Uniped, 47*(1). https://doi.org/10.18261/uniped.47.1.2

Giskeødegård, M., Wahl, A. M., & Sellberg, C. (2024). Utvikling av samarbeidskompetanse gjennom simuleringer i grunnleggende profesjonsutdanninger. I Halvorsen, Båtevik, Folkestad, Løvoll og Stokken (red) *Samarbeid Fjordantologien* 2024. Universitetsforlaget. Side 290–307.

Giskeødegård, M. F., & Wahl, A. M. (2025). Signature pedagogies of Norwegian simulator-based nautical training. *Discover Education, 4*, 148.
Grech, M. R., Horberry, T. J., & Koester, T. (2008). *Human factors in the maritime domain.* CRC Press.
Hamstra, S. J., Brydges, R., Hatala, R., Zendejas, B., & Cook, D. A. (2014). Reconsidering fidelity in simulation-based training. *Academic Medicine, 89*(3), 387–392.
Hontvedt, M., & Arnseth, H. C. (2013). On the bridge to learn: Analysing the social organization of nautical instruction in a ship simulator. *International Journal of Computer-Supported Collaborative Learning, 8*, 89–112.
Hontvedt, M., & Øvergård, K. I. (2020). Simulations at work —a framework for configuring simulation fidelity with training objectives. *Computer Supported Cooperative Work, 29*, 85–113.
Hiroaki, S., Shingo, Y., Kyoko, T., & Teppei, U. (2022). Remote use of shiphandling simulator: BRM skill acquisition. *Journal of Navigation, 75*(4), 813–831. https://doi.org/10.1017/S0373463322000352
IMCA. (2024). *Code of practice for the training and experience of key DP personnel.* IMCA Recommended practice M117 Rev. 3.2
IMO. (2011). *International Convention on Standards of Training, Certification and Watchkeeping for Seafarers. STCW. Including 2010 Manila Amendments.* International Maritime Organization.
IMO. (2014). *Leadership and Teamwork, Model Course 1.39.* Model course series by the International Maritime Organization. https://doi.org/10.62454/KT139E
IMO. (2023). *Bridge Resource Management. Model Course 1.22.* Model Courses series by the International Maritime Organization. https://doi.org/10.62454/KTB122E
Kennerley, A. (1993). Merchant Marine Education in Liverpool and the Nautical College of 1892. *The International Journal of Maritime History*. https://doi.org/10.1177/084387149300500239
Kennerley, A. (2000). Early state support of vocational education: The department of Science and art navigation schools, 1853–63. *Journal of Vocational Education & Training, 52*(2), 211–224.
Kim, T., Sharma, A., Bustgaard, M., Gyldensten, W. C., Nymoen, O. K., Tusher, H. M., & Nazir, S. (2021). The continuum of simulator-based maritime training and education. *WMU Journal of Maritime Affairs, 20.*
Lindøe J. O. (2009). *Inn fra havet. Bøyelasternes historie.* Wigestrand.
Liu, D., Macchiarella, N. D., & Vincenzi, D. A. (2009). Simulator fidelity. In D. A. Vincenzi, J. A. Wise, A. Mouloua, & P. A. Hancock (Eds.), *Human factors in simulation and training* (pp. 61–73). CRC Press.
Manuel, M. E. (2017). Vocational and academic approaches to maritime education and training (MET): Trends, challenges, and opportunities. *WMU Journal of Maritime Affairs, 16*(3), 473–483.
McLain, M. (2022). Towards a signature pedagogy for design and technology education: A literature review. *International Journal of Technology and Design Education, 32*(3), 1629–1648.
Nautical Institute. https://www.nautinst.org/technical-resources/technical-library/dynamic-positioning/ni-dp-scheme.html Retrieved 19.04.2025
Praetorius, G., Hult, C., & Österman, C. (2020). Maritime resource management: Current training approaches and potential improvements. *TransNav, 14*(3), 573–584.
Quick, J. H. (1961). The radar simulator and collision avoidance at sea. *Journal of Navigation, 14*(1), 64–68.
Rasmussen, J. (1997). Risk management in a dynamic society. *Safety Science, 27*, 183–213.
Reason, J. (1997). *Managing the risks of organizational accidents*. Ashgate.
Rystedt, H., & Sjöblom, B. (2012). Realism, authenticity, and learning in healthcare simulations: Rules of relevance and irrelevance as interactive achievements. *Instructional Science, 40*, 785–798.
Salas, E., Bowers, C. A., & Rhodenizer, l. (1998). It is not how much you have but how you use it: Toward a rational use of simulation to support aviation training. *The International Journal of Aviation Psychology, 8*(3), 197–208.

Schein, E. H. (1996). Kurt Lewin's change theory in the field and in the classroom: Notes toward a model of managed learning. *Systemic Practice and Action Research, 9*(1), 27–47.
Shulman, L. S. (2005). Signature pedagogies in the professions. *Daedalus, 134*(3), 52–59.
Thomas, L., & Reeves, S. (2015). Sociological fidelity: Keeping the patient at the heart of interprofessional learning. *Journal of Interprofessional Care, 29*(3), 177–178.
Türkistanli, T. T. (2024). Advanced learning methods in maritime education and training: A bibliometric analysis on the digitalization of education and modern trends. *Computer Applications in Engineering Education, 32*(1), e22690.
Tusher, H. M., Munim, Z. H., & Nazir, S. (2024). An evaluation of maritime simulators from technical, instructional, and organizational perspectives: A hybrid multi-criteria decision-making approach. *WMU Journal of Maritime Affairs, 23*, 165–194.
Wahl, A. M. (2020). Expanding the concept of simulator fidelity: the use of technology and collaborative activities in training maritime officers. *Cognition, Technology & Work*, 209–222. https://doi.org/10.1007/s10111-019-00549-4
Wahl, A. M., Giskeødegård, M., & Sellberg, C. (2025). Routinization, ritualization and role-play: Comparing nursing, nautical and bioengineering students' perceptions of simulation-based training. *International Journal of Training and Development, 29*, 318–327.
Wahl, A. M., & Kongsvik, T. (2018). Crew resource management training in the maritime industry: A literature review. *WMU Journal of Maritime Affairs, 17*, 377–396.
Wahl, A. M., Kongsvik, T., & Antonsen, S. (2020). Balancing Safety I and Safety II: Learning to manage performance variability at sea using simulator-based training. *Reliability Engineering & System Safety., 195.*
Weekes, M. W. M. (1962). The use of simulators for radar training. *Journal of Navigation, 15*(3), 262–267.
Wiig, A. C., Sellberg, C., & Solberg, M. (2023). Reviewing simulator-based training and assessment in maritime education: A topic modelling approach for tracing conceptual developments. *WMU Journal of Maritime Affairs, 22*, 143–164.

Aud Marit Wahl is an Associate Professor in Technology Management at the Department of Industrial Economics and Technology Management, with the Norwegian University of Science and Technology (NTNU). She is a certified commercial pilot, holds a master's degree in sociology, and a PhD in safety leadership. She has a broad experience from various industries, including aviation, railways, the maritime sector, and applied research. Her scientific work is characterized by an interdisciplinary approach and includes expertise related to simulator-based training, civil security, organizational safety, and leadership in high-risk industries.

Marte Fanneløb Giskeødegård holds a PhD in social anthropology, with specialization in organization, work, and technology. She currently works as an Associate Professor at the Department of Ocean Operations and Civil Engineering at the Norwegian University of Science and Technology. She was the project leader of a research project funded by the Norwegian Research Council called Professional education and simulation-based training (PROSIM) that compared how different professional educational programs use simulation-based training as an educational method. Her work focuses in particular on interaction and coordination in work processes, emphasizing the sociomaterial character of work.

Designing Simulation for Learning

Susan Harrington

Abstract This chapter explores the critical elements of connecting learning objectives to practical tasks within educational curricula, establishing a strong connection between the practical task and course objectives. The chapter emphasises the importance of selecting the appropriate simulator for the task at hand, determining an appropriate level of difficulty for the students, and implementing relevant briefing-scenario-debriefing designs to advance the educational benefits of simulation-based learning. Four-Component Instructional Design is introduced as a model for systematically designing simulations for learning, representing both a proven approach to designing effective instruction, and a formalisation of current scenario design practice within MET.

Keywords Scenario design · Instructional design · Four-component instructional design (4CID) · Simulation-based training · Maritime education and training (MET)

1 Introduction

Within maritime education and training (MET), simulation features prominently as part of an educational curriculum aiming to equip students with the knowledge, skills, and competencies required for a successful seafaring career. In training as future seafarers who can be relied upon to act quickly and effectively should an unexpected event occur at sea, students must develop non-technical skills such as

S. Harrington (✉)
Department of Applied Information Technology, University of Gothenburg, Gothenburg, Sweden

Department of Computer and Information Science, Linköping University, Linköping, Sweden
e-mail: susan.harrington@ait.gu.se; susan.harrington@liu.se

C. Sellberg (ed.), *Simulations in Maritime Education and Training*, WMU Studies in Maritime Affairs 12, https://doi.org/10.1007/978-3-032-08394-4_2

collaboration, communication, and decision making, as well as learning core theoretical material, and understanding the various rules and regulations applicable to their future role. Simulators provide a safe environment where students can practice their skills, experiencing the complexity of safe operations on board a ship, and building an understanding of how the various knowledge, skills, and attitudes introduced during their education are brought together as they perform the tasks required to complete an assigned scenario. While it will be some time before, for example, a second-year student takes a position of responsibility, simulation provides invaluable experiential learning through allowing them to assume the role of an expert, and how crucial it is to be proactive in tackling unforeseen events. Instructors tasked with designing scenarios must craft them carefully, connecting the learning objectives of the course with the practical tasks to be performed during the exercise, and doing so in a way that allows students to develop themselves as reflective practitioners.

Despite the value of simulation being recognised, there remains few resources for instructors wishing to learn more about designing effective scenarios, beyond utilising their own professional experience. Of course, this experience is invaluable, and the systematic approach outlined in this chapter prioritises protecting this element of scenario design, while enhancing its benefits through drawing upon instructional design models from the learning sciences. Some 13 years after the need for a systematic approach to scenario was highlighted by Baldauf et al. (2012), instructors participating in an interview study around scenario design practice confirm a need for resources to assist them in designing scenarios conducive to learning (Harrington et al., 2025). As technology becomes increasingly more advanced, it becomes ever more important to remember the human element in simulator-based MET, ensuring that the strengths of learning from professional experiences are preserved, and students benefit from instructors' expertise. Even the most advanced simulator is simply a tool, only effective when utilised appropriately.

When the term 'scenario design' is used in this chapter, it encompasses the three phases of simulation—briefing, scenario, and debriefing—as seen in Fig. 1, although briefing and debriefing are discussed in more detail in chapters "Preparing for Simulation Through Briefing" and "Facilitating debriefings for feedback, discussion, and reflection" respectively. It also considers that simulation is not a standalone activity, but rather an opportunity for students to put into practice what they have learned to date, meaning that instructors engaging in scenario design should consider the training program holistically, making design decisions accordingly. The chapter begins by introducing current approaches to scenario design, drawing upon accounts provided by experienced instructors, before moving on to discuss instructional design, the suitability of a specific instructional design model in this context, and, finally, outlining a framework for systematic scenario design.

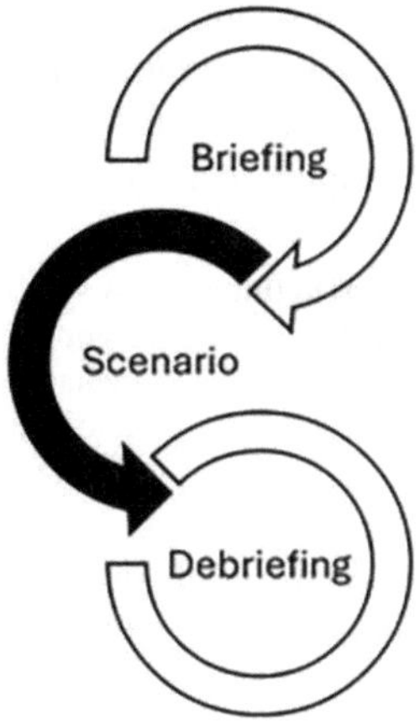

Fig. 1 Briefing-scenario-debriefing model

2 Designing Simulation for MET

While simulation is used for a variety of purposes within MET, the overarching objective is to train future professionals, competent in working effectively and safely in a high-risk environment (e.g., Sellberg, 2018). To paraphrase one instructor, a student is ready to graduate when the instructor feels they could sleep on a ship where the student is in charge.

Traditionally, this journey from novice with absolutely no knowledge and experience, to expert who could competently take charge of the ship took place at sea, in the guise of a master-apprentice relationship. Here, novices learned from experienced mariners, gaining both explicit and tacit knowledge through working alongside experts over an extended period of time. As the novice progressed, the difficulty of tasks and level of responsibility increased. These practices naturally led to the kind of holistic learning that is required—through being immersed in life on board a ship, practical tasks were mastered through practice and repetition, but so too were the social elements of being a seafarer, for example, expected ways of behaving, communicating, and working as part of a team. While this model was not without its drawbacks, this very simplified version serves as a useful illustration of the overarching objective of MET being to develop students as future seafarers. Instructors bring a wealth of experience to their teaching; however, simulator time is limited and expensive. Effective simulation for learning relies upon careful design of scenarios that facilitate this transfer of knowledge from instructor to student, and from student to their own professional practice.

Complex learning	Learning which involves the integration of knowledge, skills, and attitudes through coordinating different constituent skills in a way that allows transfer of what is learned during education to professional practice (Kirschner & van Merriënboer, 2008)

This chapter aims to provide a resource for instructors to use when designing scenarios, addressing the need for a systematic approach to scenario design that combines instructors' experience with an instructional design model developed for complex learning.

Instructional design	"A systematic approach to the design, production, evaluation, and utilization of complete systems of instruction, including all appropriate components and a management pattern for using them" (Association for Educational Communications and Technology, 1977 p. 172 cited in Branch & Kopcha, 2014, p. 78)

In the context of scenario design, the definitions of complex learning and instructional design provided here can be interpreted as designing scenarios that facilitate students' training as future professionals, encouraging self-directed learning. Instructors' expert knowledge is elicited through following the steps of 4C/ID, which also considers all other elements of the educational program (lectures, class materials, cadetships), regulations which students must learn to comply with, technical and non-technical skills. Instructors using 4C/ID are prompted to set clear learning objectives, tailor the scenario to students' current level of knowledge and experience, and treat scenario design as part of a holistic design process which promotes the transfer of learning to professional practice.

Holistic design	Design which considers all elements of complex learning, and their interconnected nature. This avoids the problems of compartmentalization (separating learning into constituent parts), fragmentation (breaking learning down into too-small, or incomplete parts), and the transfer paradox (focusing on efficiency rather than on the construction of knowledge that can be applied in future, as-yet-unknown situations) (Kirschner & van Merriënboer, 2008)

As such, when working to design simulation for effective learning in MET, the primary concern should not be training a specific skill, but on how to incorporate that skill into the students' training as future competent seafarers. Real-life experiences and events offer excellent opportunities for training, but instructors designing simulator exercises should bear in mind that students are unlikely to face that exact event in their own professional practice. However, they are likely to face other unforeseen events, where they must draw upon their repertoire of knowledge and skills to avoid potentially catastrophic consequences. Each of the instructors participating in the study conducted by Harrington et al. (2025) expressed this sentiment, emphasising its key role in their own scenario design practices. The 4C/ID instructional design model has been selected due to its close similarity with the current practices described by instructors participating in a series of semi-structured interviews on scenario design (Harrington et al., 2025), thus offering a means of formalising and systematising proven current practices. 4C/ID, as laid out in this chapter, is proposed as a solution to the lack of scenario design resources which was expressed as a common concern amongst the participants in the study.

3 The Challenges of Designing Simulation for Learning

Several challenges present when designing scenarios for simulation training, some specific to the domain of MET, and others more generally applicable. The lack of a formalised, systematic approach to scenario design in MET is one, as is the need to take into consideration the social aspects of training to be a competent seafarer. The specifics of the nature of MET are beyond the scope of this chapter but are discussed further elsewhere.

In Harrington et al. (2025), a series of semi-structured interviews with simulator instructors was used to explore the core considerations and challenges of scenario design. These were summarised under three key themes—functional congruence, tailoring learning to students, and clear learning objectives.

Functional congruence is a term used by the authors to build upon the use of realism, authenticity, fidelity, and functional resemblance when discussing scenario design, described by one instructor as the "*clear and distinct link between what you do in the simulator, and what could happen, or has happened, on board*". Functional congruence involves striking an often very delicate balance between a realistic simulated environment, in terms of technology, sounds, communication, stress levels and similar elements, and the main educational objectives of the training. For effective learning, students should not be either overwhelmed or bored, but must also encounter realistic challenges and obstacles, such as poor visibility in an ice navigation scenario, while acting in accordance with the norms of bridge operations.

Designing for students involves tailoring the scenario to the student's level—the first scenario encountered by students will be very different to scenarios designed for final-year students, in terms of content and complexity—both when designing the scenario and monitoring and assessing performance. A first-year student may encounter a scenario where they are only required to set up the instruments on the bridge, whereas a final-year student is required to operate alone during a navigation scenario, demonstrating their competence. Instructors adjust their role in the scenario according to the students' actions, with exercises becoming more and more advanced over time, requiring students to use knowledge gained throughout their course. This must be done in a way where there is a balance between challenging the students and instilling them with confidence in their own ability.

Setting clear learning objectives is a key element of effective scenario design. Students cannot achieve an objective if it is not clear, and instructors cannot design a scenario which enables them to achieve an objective unless they are certain about what the objective is. One example provided during the series of interviews was that safe navigation may be a great aim to have, but it is not a clear learning objective for which a scenario can be designed. However, "*navigating according to COLREG 7 and 8*" is. When using the guidelines provided later in this chapter, instructors are prompted to formulate learning objectives in this way. Closely aligned with this emphasis on learning objectives is the STCW Convention statement that "*any simulator used for mandatory simulator-based training must be suitable for the selected objectives and training tasks*".

The 4C/ID instructional design model offers a framework for addressing these challenges, alongside related challenges of designing effective simulation for learning. These challenges are briefly described here as an illustration of why this particular model was deemed appropriate in the context of MET.

The Complex Nature of the Tasks Involved Students engaging in simulator-based MET are working towards, or developing within, a profession where risk is high, and they must be able to act appropriately in a range of potential circumstances. As such, the tasks involved in simulator-based training can be considered complex tasks, where a complex task is one where performance requires both the automated application of knowledge, and more controlled, deliberate application of explicit knowledge (Clark et al., 2008), or, in learning terms, one which requires the integration of knowledge, skills and attitudes (Kirschner & van Merriënboer, 2008; van Merriënboer et al., 2006). In other words, as students perform simulated scenarios, the aim is generally not for them to follow a prescribed course of action, but to demonstrate that they can work effectively and successfully as part of a team. There is not always a single correct outcome or action that the students should aim for, but instructors rely on their experience and professional intuition in assessing factors such as decision-making, collaboration, and communication. While this experience and intuition allows instructors to evaluate student performance, they are not always aware of the criteria they use, or the reasoning behind their own decision-making during assessment (see chapter "Assessing Competence Through Simulation" for more on assessment). Similarly, extensive experience means that instructors are not always consciously aware of every step or piece of necessary information required by students to perform a certain task, due to having performed that task so many times during their career that they can now do it automatically.

Compartmentalisation and Fragmentation of Tasks (Frerejean et al., 2019) Where complex tasks are involved, effective teaching and learning requires more than treating each task or instruction as standalone knowledge. For example, using the instruments on the bridge during navigation training is just part of the overall objective, which goes far beyond the students' interaction with each of the instruments. While it is essential for a student to know how to correctly use the various instruments available to them, bringing this knowledge into use as part of handling the overall scenario being simulated is more important. As such, when designing scenarios, it is paramount that exercises are designed in a way which facilitates students in developing their repertoire of knowledge, skills, and attitudes through bringing together what they have learned to date. A whole-task approach, and integration of elements is more appropriate here, and avoids the transfer paradox (Kirschner & van Merriënboer, 2008) where sequencing of tasks as standalone elements of increasing difficulty sacrifices effective learning in favour of efficiency in terms of time and effort. This comes back to the earlier point about developing students as future seafarers—by focusing on promoting overall reflective practice, rather than seeking the quickest route to training a single task, this potential issue can be avoided.

Suitability of Tasks Considering the Learner's Existing Knowledge Design must take into account what the learner already knows, and what is likely to be feasible for effective learning in a simulator exercise given this prior knowledge. Various factors including working memory capacity, the language used, and stress levels impact upon how suitable a task may be (Baldauf et al., 2012). Cognitive load theory can be used to help with balancing intrinsic, extraneous and germane load for more effective learning (Sweller, 2010; van Merriënboer et al., 2006; van Merriënboer & Sluijsmans, 2009). In using 4C/ID to guide scenario design, cognitive load is taken into consideration, being built into the model itself. Thus, using the model for scenario design helps to ensure that tasks are designed at an appropriate level for the students in question, without requiring in-depth understanding of cognitive load theory on the part of the instructor. This allows instructors—and the students performing the scenario—to benefit from scientific knowledge on effective design of learning activities, and effectively communicating their professional expertise through clear learning objectives and appropriately-structured tasks.

Elicitation of the Instructor's Expert Knowledge The importance of communicating professional expertise as part of a process of developing students as future seafarers brings us to the next challenge facing those engaging in scenario design. Experts, in this case the instructors, can carry out tasks with much more ease than a novice, as elements of their actions will have become automated through practice. As experts are not always aware that they are carrying out these automated actions, let alone able to express them clearly to students, teaching and learning which relies on the experts' accounts of how a task is performed can miss some 70% of the information that is important for learners' development (Clark, 2014). Effective scenario design must take an approach that allows for the elicitation of this knowledge. Similarly to how it is important to understand what is meant by "*good enough*" when assessing student performance, it is also essential to ensure that students are being provided with all necessary information.

Why 4C/ID?	• Similar to current practice, but offers a much-needed resource for guiding and structuring scenario design • Proven effective for designing complex learning • Facilitates the development of a repertoire of knowledge, skills, and attitude transferrable to professional practice, i.e., in developing students as future professionals

Using the 4C/ID model to structure scenario design forces instructors to actively consider every step of performing the task, and the criteria they have in mind while evaluating performance. This move from "*knowing the scenario is good enough*" to determining exactly what "*good enough*" means, enhances learning by ensuring the objectives and evaluation criteria are clear for both students and instructors, and by capturing the instructors' tacit knowledge in a way which can be explicitly communicated to students. In turn, this addresses the concern raised by Baldauf et al.

(2012) around the effectiveness of taking an event-driven approach to scenario design, where scenario design is based primarily on past experiences and events. As highlighted by the authors, although simulator technology is advancing, and regulations are updated, accidents at sea continue to occur.

The 4C/ID model lends itself well to the current practice within MET, where instructors follow an input—process—output model in the form of briefing—scenario—debriefing. 4C/ID is intended to provide more structure to the process itself, by systematizing how instructors approach the design of the scenario. As simulator-based training in MET involves students working through scenarios based on real skills and competencies they will require in their professional lives, and mirroring real-life, or at least very realistic, events and situations they could feasibly encounter. An appropriate instructional design model, therefore, is one that allows the instructor to ensure that all relevant elements of the task in question have been captured and incorporated in the scenario exercise, working as a complementary addition to the tried-and-tested briefing—scenario—debriefing model. As such, the proposed approach embraces the strengths of intuitive instructional design based on extensive professional experience, while also reaping the benefits of a systematic approach (Hoogveld et al., 2002, 2005), such as facilitating instructional design that is tailored to both educational and professional required outcomes, and student knowledge and needs. That is, scenarios that are purposefully and carefully designed to replicate professional practice, while also founded upon strong pedagogical principles. Taking a systematic approach also encourages instructors to consider best practices, student feedback on previous simulations, and their own experiences on what did or did not work well in the past.

While selecting an appropriate instructional design model for this purpose, various elements of MET were taken into consideration, to ensure that the framework for scenario design complements the nature of education and training in the MET field and preserves the benefits of the current approach. These elements include:

- Recognising the instructor's own experience as being an invaluable resource, and drawing upon it. The inclusion of real-life practice is essential when using the 4C/ID model.
- Social learning from peers, and from reflecting upon one's own performance both during and after simulation e.g., during debriefing.

The use of 4C/ID represents a move to a learning objective-based approach, focusing on utilising learning theory for effective education of competent seafarers. Through using this model to guide scenario design, instructors can relate regulations to concrete learning objectives, tailor complexity to students' existing level of knowledge, and develop exemplars for evaluation and assessment. Such an approach to scenario design also encourages instructors to think about the potentials and limitations of the simulator itself, in terms of how the learning objective is to be achieved (see Harrington et al., 2025 for more detail on this aspect of scenario design).

4 Using 4C/ID in Simulation Design

Now that we have considered the elements of scenario design in MET, and the challenges associated with designing simulation for learning, the next question is *how do we actually design effective scenarios*?

The final section in this chapter provides guidelines for implementing the 4C/ID model in scenario design, adapting the 10 steps of 4C/ID (Kirschner & van Merriënboer, 2008) to the MET context. While these guidelines may initially seem daunting, bear in mind that 4C/ID was developed specifically for use by those less experienced in instructional design, and is deliberately detailed to ensure that no essential information or elements of instruction are missed. An illustration of the process is provided at the end of the chapter (see Fig. 2).

Note that the aim here is not to design scenarios as quickly as possible, but to design effective scenarios whose performance aids in equipping students with the knowledge, skills and competencies required for successful professional practice. While approaching the scenario design process in this task-centred way requires more initial time investment than an intuitive approach to scenario design, the integrative nature of the resulting learning makes it more efficient and effective in the longer term, achieving the goal of developing students as reflective professionals, competent in handling unforeseen and complex events. Of course, implementing the approach will also become less time consuming with practice.

4.1 The 4 Components of 4C/ID

Implementing the 4C/ID approach involves following the design process and design principles embedded within the model (van Merriënboer, 2021). These cover elements such as setting standards for acceptable performance, sequencing learning tasks, and designing supportive and procedural information. That is, defining criteria for assessment of students in the simulator, deciding upon the appropriate order in which certain tasks and competencies should be introduced and trained, and ensuring that students are provided with necessary information in lectures and teaching materials prior to the simulator session, and also during simulation as appropriate. Table 1 provides a brief overview of the 4 components in the MET context.

4.2 Implementing 4C/ID for Scenario Design

Firstly, before embarking upon the guidelines for implementing 4C/ID for scenario design, it seems important to acknowledge the apparent complexity of the process. It is necessary to provide a lot of information at this point due to the assumption that

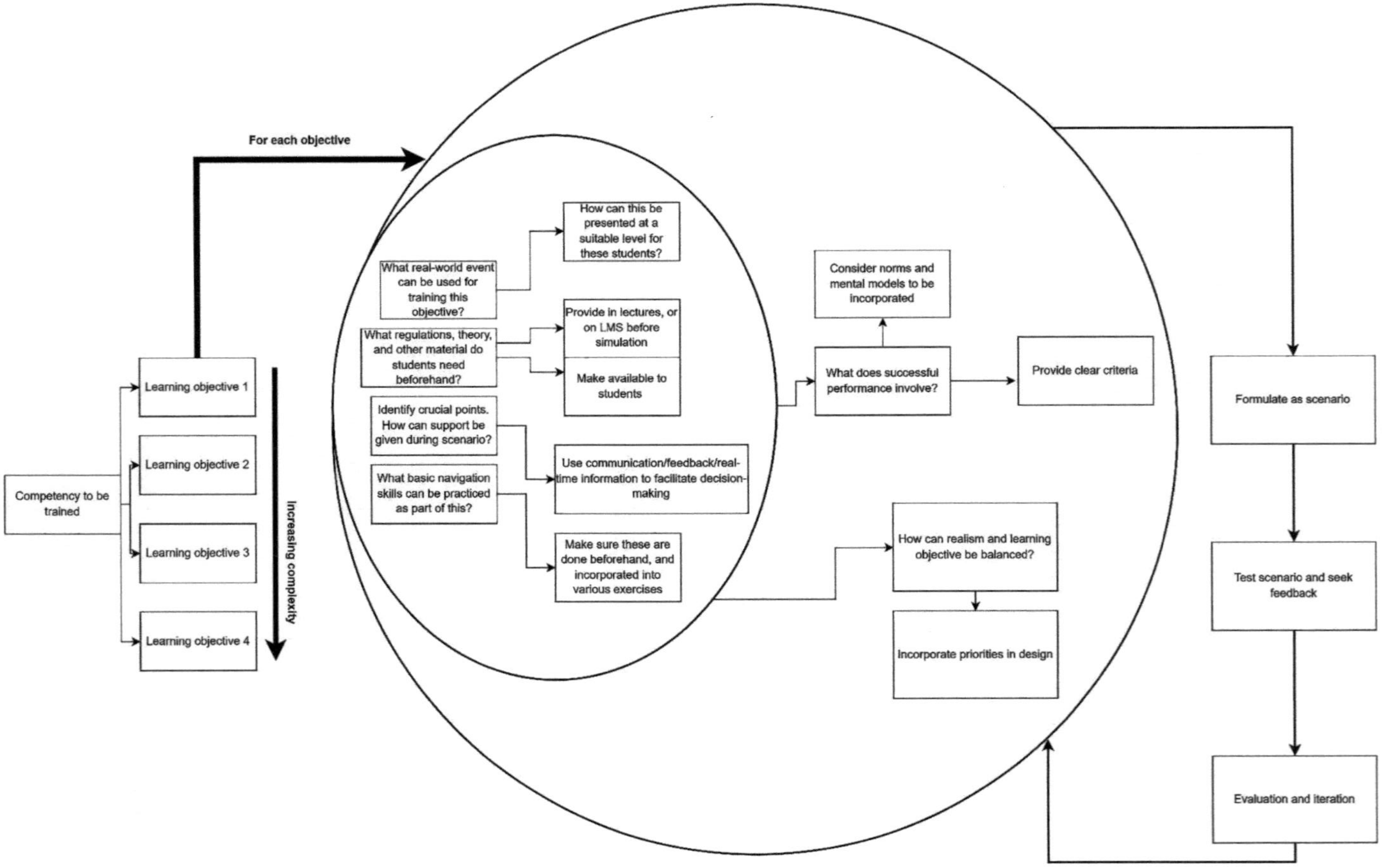

Fig. 2 Scenario Design Process

Table 1 Applying 4C/ID to the MET Context (Harrington, 2024)

Learning Tasks	Identify the whole task to be trained. This should be a task or competency that students can expect to perform as a maritime professional e.g., navigation, ship-handling, communication, or emergency response. Break the task down into manageable chunks, creating clear, measurable, learning objectives. Draw upon real-world events requiring the performance of these tasks. The scenario should be challenging enough to be engaging without being overwhelming, and should provide learners with opportunities for practicing constituent skills
Supportive Information	Provide students with background information supporting the tasks, e.g., regulations, relevant theoretical concepts. This should be provided before students attend the simulator, e.g., in a lecture, or in text or multimedia format on a virtual learning environment. Ensure that this information is integrated into the scenario to facilitate students increasing understanding in context.
Procedural(just-in-time) Information	Expert knowledge and real-world experience should be used to identify critical points in the scenario where students may require guidance or additional information Provide clear, concise information at these critical points, to assist students in making informed decisions or successfully handling challenging events This information should be made easy for students to access when needed
Part-Task Practice	Break whole tasks into smaller components (part-tasks) that students can practice individually before integrating them into the practice of the whole complex task. This facilitates the automation of part-tasks that will be required regularly in professional practice, *where the whole task does not allow for sufficient practice*. The part-tasks should first be encountered as the whole task, before being introduced for individual practice. Subsequent scenarios should include repetition of these part-tasks, increasing in complexity to allow students to gradually improve their skills. Feedback should be provided during part-task practice to correct any errors and encourage good practice.

those designing scenarios for simulator training are not familiar with the model, however, please rest assured that it is not as complicated as it may seem at first. In addition to the detailed guide (adapted from van Merriënboer, 2021) which follows, an illustrated overview of the design process is provided for reference (see Fig. 2).

As the steps have been presented in a general way, it may be helpful to have a specific course or objective in mind when reading through the guide.

5 Step 1: Design Learning Tasks (Objectives)

The first step in the process of implementing 4C/ID involves designing the learning tasks. These relate to the objectives of the simulation training, bearing in mind the importance of considering the course holistically. Scenarios should build upon what students have learned previously, and provide them with the opportunity to practice in a safe, controlled environment congruent with reality.

- Tasks should be based upon those encountered when operating in the role students are simulating, and should require students to draw upon their bank of knowledge, skills, and attitudes. For example, navigating according to COLREG 7 and 8
- As the aim is for students to progress gradually from being novice students to becoming competent mariners, the difficulty of the tasks and learning objectives should increase in line with student ability, in terms of both difficulty and fidelity. For example, navigating according to a predefined route, followed by completing parts of the preparation themselves, before finally taking responsibility for all planning, preparation, and execution
- Where possible, the scenario should include all elements of the corresponding real-world event. For example, emergency calls from other vessels, communication from VTS, use of radar and ECDIS, communication between engine room and bridge
- The level of support provided to students should be appropriate to their level of knowledge and experience. This may include demonstrations, examples, and similar instructor support in performing the exercise
- Students should be provided with information on norms, expected practice, and relevant information where appropriate. The timing of receiving this information is determined by whether it is considered supportive information (theory) or procedural (just-in-time) information
- Support and guidance should decrease over time, in line with students' increasing ability. The ultimate aim is for students to work independently

6 Step 2: Set Performance Standards and Objectives

Once the learning tasks (objectives) have been designed, the next step in the scenario design process involves determining the standards of acceptable performance, performance objectives for each skill involved in performing the learning task, and clear criteria for each.

- All skills and actions necessary for performing the task designed in Step 1 should be mapped, with instructors taking care to consider all elements (including those which they, as experts, now perform automatically). This map then serves as an overview to be used when assessing student performance
- A performance objective should be set for each of the mapped skills, comprising expected standards (including expected use of resources during simulation). Generally, a clear performance objective should use an action verb, e.g. monitor, communicate, demonstrate…
- Each objective should be classified as either routine (requiring just-in-time information), non-routine (requiring supportive information), or automated (requiring part-task practice)

- Clearly define the acceptable standard for performance of each objective. This should include criteria for assessing performance, e.g., time taken, accuracy, compliance with appropriate regulations, and elements such as expected communication standards
- Develop a scoring rubric for each of the standards, enabling assessment of student performance in the simulator, and monitoring of student progress over time

7 Step 3: Arrange Learning Tasks in Order of Increasing Complexity

The next step in the process involves ensuring that tasks are presented in order of increasing complexity, tailored to individual student needs, and encouraging self-directed learning where appropriate in guiding students towards becoming competent professionals.

- Tasks should increase in difficulty from the simplest to the most complex. While this sequencing may be straightforward in some cases, it may also be necessary to consider what conditions and factors make a task simple or complex, in order to ensure appropriate sequencing
- Where appropriate, decreasing support can be used to increase complexity. For example, students may first be presented with a prepared navigation route, then tasked with adding to a partially prepared plan, before finally progressing to complete all preparation independently
- The level of complexity, and the support and guidance provided, should be appropriate for the students' needs
- Self-directed learning, where students are given some control over choosing learning tasks, can be used as a form of scaffolding where responsibility is gradually passed from instructor to student

8 Step 4: Provide Supportive Information for Non-routine Tasks

This step involves ensuring that students are provided with the supportive information, or theory, needed for performance of the learning tasks. This may include manuals, regulations, information on expected norms and similar, and should be provided in advance of simulation e.g., in a lecture or on the course learning management system.

- Supportive information includes all information that students need in order to be able to successfully perform the non-routine elements of learning tasks, e.g.,

regulations, handbooks, approaches to problem solving, norms around how things are done in the maritime industry
- Students should be provided with information on any conceptual models, norms, or other relevant information on how the maritime domain is organised, that will aid them in acting as expected of a professional seafarer
- Where there are standard approaches to problem-solving, or descriptions of the phases of performing a specific task, these should be provided to students
- Students should be provided with feedback, e.g., in debriefing, that is formulated in way that encourages them to reflect upon their own performance, mental models, and strategies, and compare these to those of instructors, peers, or other experts (see chapter "Facilitating debriefings for feedback, discussion, and reflection" for more on debriefing)

9 Step 5: Provide Just-in-Time Information and Part-Task Practice of Routine Tasks

Procedural information, or just-in-time, information is that which should be provided to students as needed, rather than in advance. This information relates to routine aspects of performance and informs students on the "how-to" of performing tasks. For example, this may include instructors communicating as VTS during a scenario—students would typically not be told this is going to happen at a specific point, but the call would be received at an appropriate juncture

- Procedural information includes the "how-to" information and corrective feedback that helps students to perform routine elements of the learning task
- How-to instructions are provided just-in-time to inform students on how to perform routine tasks. It may be immediately obvious what information students will need and when they will need it, but instructors designing scenarios may need to actively consider this. As experts, these routine tasks will be performed automatically, and may need some effort to identify all constituent parts. Any prerequisite knowledge must also be considered
- For routine tasks, feedback should be provided immediately where students make an error e.g., making a call as VTS, asking relevant questions and providing instructions that indicate to students that they have not acted as expected. Feedback should explain the cause of the error, and hints on how to recover from the error while continuing the task
- Part-task practice should be integrated into the design of scenarios to help students in fully automating routine tasks. This practice should focus on accuracy first, then speed, then occurring at the same time as other tasks. Note that part-task practice may not be appropriate where it is difficult to separate elements of a task, but care should still be taken to provide sufficient opportunity to practice routine tasks

Once these 5 steps have been completed, the newly-designed scenario should be tested and refined in an iterative process of designing effective simulation for learning.

The next chapter in this book, “Preparing for Simulation Through Briefing”, discusses preparing for simulation through pre-briefing.

References

Baldauf, M., Nolte-Schuster, B., & Schröder-Hinrichs, J.-U. (2012). A systematic approach for simulation-based team training in maritime safety. In *INSLC 17 - international navigation simulator lecturers' conference*. Rostock-Warnemüde.

Branch, R. M., & Kopcha, T. J. (2014). Instructional design models. In J. M. Spector, M. D. Merrill, J. Elen, & M. J. Bishop (Eds.), *Handbook of research on educational communications and technology* (pp. 77–87). Springer. https://doi.org/10.1007/978-1-4614-3185-5_7

Clark, R. (2014). Cognitive task analysis for expert-based instruction in healthcare. In J. M. Spector, M. D. Merrill, J. Elen, & M. J. Bishop (Eds.), *Handbook of research on educational communications and technology* (pp. 541–551). Springer. https://doi.org/10.1007/978-1-4614-3185-5_42

Clark, R., Feldon, D. F., Van Merrienboer, J. J. G., Yates, K., & Early, S. (2008). Cognitive task analysis. In *Handbook of research on educational communications and technology* (pp. 577–593). Taylor and Francis Group.

Frerejean, J., van Merriënboer, J. J. G., Kirschner, P. A., Roex, A., Aertgeerts, B., & Marcellis, M. (2019). Designing instruction for complex learning: 4C/ID in higher education. *European Journal of Education, 54*(4), 513–524. https://doi.org/10.1111/ejed.12363

Harrington, S. (2024). Enhancing scenario design in maritime education and training. *Advances in Human Factors of Transportation, 148*(148) https://doi.org/10.54941/ahfe1005266

Harrington, S., Sellberg, C., & Lindwall, O. (2025). Crafting congruence: Towards systematic scenario design in maritime simulator training and assessment. *WMU Journal of Maritime Affairs*. https://doi.org/10.1007/s13437-024-00356-3

Hoogveld, A. W. M., Paas, F., & Jochems, W. M. G. (2005). Training higher education teachers for instructional design of competency-based education: Product-oriented versus process-oriented worked examples. *Teaching and Teacher Education, 21*(3), 287–297. https://doi.org/10.1016/j.tate.2005.01.002

Hoogveld, A. W. M., Paas, F., Jochems, W. M. G., & Van Merriënboer, J. J. G. (2002). Exploring teachers' instructional design practices from a systems design perspective. *Instructional Science, 30*(4), 291–305.

Kirschner, P., & van Merriënboer, J. J. G. (2008). Ten steps to complex learning: A new approach to instruction and instructional design. In T. Good (Ed.), *21st century education: A reference handbook* (p. I-244-I–253). SAGE Publications,Inc.. https://doi.org/10.4135/9781412964012.n26

Sellberg, C. (2018). From briefing, through scenario, to debriefing: The maritime instructor's work during simulator-based training. *Cognition, Technology & Work, 20*(1), 49–62. https://doi.org/10.1007/s10111-017-0446-y

Sweller, J. (2010). Element interactivity and intrinsic, extraneous, and germane cognitive load. *Educational Psychology Review, 22*(2), 123–138.

van Merriënboer, J. J. G. (2021). *The four-component instructional design model. An Overview of its Main Design Principles*. Maastricht University. https://www.4cid.org/wp-content/uploads/2021/04/vanmerrienboer-4cid-overview-of-main-design-principles-2021.pdf.

van Merriënboer, J. J. G., Kester, L., & Paas, F. (2006). Teaching complex rather than simple tasks: Balancing intrinsic and germane load to enhance transfer of learning. *Applied Cognitive Psychology, 20*(3), 343–352. https://doi.org/10.1002/acp.1250

van Merriënboer, J. J. G., & Sluijsmans, D. M. A. (2009). Toward a synthesis of cognitive load theory, four-component instructional design, and self-directed learning. *Educational Psychology Review, 21*(1), 55–66.

Susan Harrington is a visiting researcher and former postdoctoral researcher at the Department of Applied Information Technology, University of Gothenburg, specializing in simulator-based maritime education and training. With a PhD in Management Science from the University of Strathclyde, she focuses on scenario-based learning design to support immersive, experiential training in high-risk maritime contexts. Her research bridges pedagogical theory and practical application, examining how scenario exercises can cultivate competencies such as communication, collaboration, reflexive decision-making, and good seamanship through structured debriefings and instructional support. Susan's research also focuses on designing effective learning analytics for simulator-based professional learning.

Preparing for Simulation Through Briefing

Charlott Sellberg and Daniel Erntsson

Abstract Briefing serves as a foundational element in maritime navigation training, setting the stage for meaningful engagement with simulation-based scenarios. These prospective and preparatory activities aim at managing student expectations and providing them with an understanding of the learning objectives, the upcoming task, and contextual condition they will encounter during the scenario. This chapter examines the pedagogical and operational aspects of briefings, exploring how this phase serves as a starting point for bridging between abstract knowledge and knowledge-in-action. By focusing on the intricacies of preparing students for simulation, planning the briefing as well as the delivery of instructions in briefings, the chapter underscores the critical role of briefings in enabling participants to approach simulations not merely as technical exercises but as opportunities for skill development and professional growth.

Keywords Simulation-based training · Pre-simulation preparation · Pre-briefing, briefing · Maritime education and training (MET)

1 Introduction

Although briefing is widely recognized as a cornerstone of simulation-based training, it remains an understudied phase within the briefing-scenario-debriefing framework in research on maritime education and training (MET). Yet this stage presents

C. Sellberg (✉)
Department of Applied Information Technology, University of Gothenburg, Gothenburg, Sweden
e-mail: charlott.sellberg@ait.gu.se

D. Erntsson
Department of Mechanics and Maritime Sciences, Chalmers University of Technology, Göteborg, Sweden
e-mail: daniel.ernstsson@chalmers.se

C. Sellberg (ed.), *Simulations in Maritime Education and Training*, WMU Studies in Maritime Affairs 12, https://doi.org/10.1007/978-3-032-08394-4_3

several practical challenges for simulator instructors. How can we create a psychologically safe learning environment in simulations, knowing that students, as novices, will inevitably make mistakes during training? How can we manage their expectations of simulation-based training? How can we ensure they arrive to the simulator adequately prepared? And how can we deliver meaningful instruction before the scenario unfolds? While the literature on maritime simulation offers few answers, the healthcare domain provides a more substantial body of documented best practices to draw upon (McDermott et al., 2021). However, in healthcare literature, pre-simulation preparation is mostly directed towards building a psychologically safe learning environment, setting the appropriate expectations, and establishing the so-called fiction contract rather than practical information introducing learning objectives and course content. The *fiction contract* in pre-briefing, that is, already before the briefing takes place, aims to establish the simulation as a learning experience rather than a clinical scenario (Kolbe et al., 2015; Sharma et al., 2023; Somerville et al., 2023). This agreement is said to frame the simulation as a safe, supportive space for learning, encouraging participants to engage fully without the fear of real work consequences. The aim of establishing a fiction contract is to reduce anxiety and emphasize psychological safety, and it has been said to help learners focus on the objectives of the simulation and approach tasks with confidence and openness (Kolbe et al., 2015).

Ideally, simulations should facilitate a safe space for learning where participants can explore different strategies, make mistakes, and learn from them without the pressures of real-work performance (Sharma et al., 2023). In this way, the fiction contract fosters a deeper engagement with the learning process and allows participants to develop critical professional competencies (Kolbe et al., 2015). By framing simulations as opportunities for growth and skill development rather than evaluations or tests, the fiction contract encourages a culture of collaborative learning, preparing students for the challenges of their future roles as maritime professionals. In healthcare literature, the fiction contract is said to ensure that participants are adequately prepared for the simulation and aware of the expectations and ground rules for the simulation experience (El Hussein et al., 2021; Kolbe et al., 2015; McDermott et al., 2021). Overall, within the healthcare literature, well-conducted pre-brief activities are said to lay the foundation for a positive and effective learning environment, where participants can engage in simulation in a psychologically safe way (Somerville et al., 2023). To foster a psychologically safe learning environment, facilitators are encouraged to establish the simulation as a safe space during the pre-briefing and briefing phases. Key practices include clarifying expectations, outlining the simulation process, defining roles, ensuring confidentiality, and creating a non-threatening atmosphere. Additional strategies such as offering tailored support, reinforcing prior knowledge, promoting respectful interactions, and organizing learners into small groups are also important for maintaining a sense of safety (ref). However, while supportive environments are essential, we argue that there is a risk in idealizing psychological safety in ways that obscure the inherently evaluative and performative aspects of professional training. Peer comparison and social exposure are often unavoidable, and it is frequently the fear of failing in front of peers rather than instructor feedback that provokes the most anxiety. Thus, psychological safety in simulation should not be understood as the elimination of discomfort, since discomfort may be both inevitable and pedagogically valuable. Instead, psychological safety means creating conditions in which learners feel supported in managing that discomfort. It involves transparent and fair expectations, consistent and respectful treatment, and accountability for preparation and prior knowledge. Rather than shielding learners from all forms of challenge, psychological safety ensures that challenges are encountered within a framework of fairness, clarity, and trust.

Moreover, even if psychological safety is an important aspect in learning from simulations, there are also pedagogically important questions that deserves further attention, i.e., how the instructor could present the scenario's learning objectives and assign different roles, how to offer an orientation to the simulated environment, and communicate key information relevant to the specific scenario (Sellberg, 2018). In this chapter, we aim to provide an overview of guidelines addressing both briefing preparation (so-called pre-debriefing) and briefing components, as well as examining how briefings are shaped by temporal, social, and material conditions of simulation-based training. The guidelines in this chapter are based on our explorations of previous research on briefing in simulation, analyses of video-recorded debriefings in basic navigation training for master mariner students and experienced professionals, as well as experiences from experienced simulator instructors.

2 Preparing Students for Simulations Through Pre-Debriefings

As seen in the introduction, it is important to set the right expectations for learning. In the literature on healthcare simulations, we have seen how this is formulated as the fiction contract. In educational research, the notion of *ground rules* refers both to the implicit or explicit shared norms and principles that govern how people interact and communicate within a group (Mercer, 2019). These rules shape the quality of dialogue, collaboration, and learning by defining what kinds of contributions are appropriate, how they should be made, and how group members respond to each other. Ground rules are particularly significant in educational and professional contexts where communication and collaboration are essential, such as in simulation-based training. Mercer (2019) emphasizes the importance of dialogue that is open, reasoned, and respectful for participants to learn from collaboration. Clarifying the ground rules for students support a positive learning environment by encouraging participants to share their ideas, challenge each other constructively, and build shared understanding. Formulating ground rules for simulation-based training need to put emphasis of the simulation as a safe space for professional learning, active engagement in the learning activities and an open and respectful dialogue. Suggestions for ground rules for learning from simulations are outlined in Fig. 1.

Fig. 1 Suggested ground rules for learning from simulation

A way of ensuring that students are prepared for simulation is to assign them the task of creating a voyage plan. In Starup et al. (2024), an analysis of a playful competition between students in a full mission bridge simulator illustrates how planning for the race involved extensive and collaborative route development through a structured process that reflects the professional demands of maritime navigation. Working within bridge teams, the students took on the responsibility of preparing a complete passage plan using desktop simulators. Over the course of a week, they developed their plans while adhering to predefined constraints that mirror the regulatory and operational limitations they will encounter at sea. With three students on the bridge at any given time, the exercise reinforced not only the technical aspects of navigation but also the collaborative and procedural nature of voyage planning in a professional setting. A central part of such planning involves gathering and analyzing essential navigational data. Using nautical charts and digital tools, students assess water depths, tidal movements, and meteorological forecasts to ensure a safe and efficient route. Regulatory considerations, such as Traffic Separation Schemes, restricted areas, and port entry requirements, must be carefully integrated into the voyage plans. Through this process, they identify waypoints, define turning points, and anticipate potential hazards that could affect the passage. Each decision is subject to discussion within the team, encouraging students to evaluate alternatives, justify their choices, and refine their approach. The ability to coordinate and critically assess different planning strategies is particularly important in high-stakes situations, such as the simulator-based competition examined in Starup et al. (2024) where teams had to balance strategic decision-making with adherence to safety regulations.

While the students in this study were in their senior year, it is important to recognize that preparing for simulation differs significantly for novices. Initially, beginners focus on the fundamental task of transferring waypoints onto a chart, ensuring that their course follows a safe and logical path, adhering to the COLREGS (Sellberg & Sharma, 2024). At this stage, planning is largely procedural, emphasizing the mechanics of using nautical charts, digital navigation systems, and standardized conventions for marking a route. The primary concern is to establish a clear passage without running into obvious hazards, but students have yet to develop a deeper awareness of the uncertainties and contingencies involved in real-world navigation. Voyage planning evolves as students progress from simple familiarization exercises and basic route plotting to more advanced decision-making that incorporates multiple layers of navigational, environmental, and operational complexity into their plans. As their experience grows, students begin incorporating additional elements into their planning, recognizing that voyage planning is not just about determining a route. It now becomes a matter of continuously assessing and adjusting to changing conditions as well as different protocols, such as, for example, when students are training for Search and Rescue (SAR) Operations. In such operations, students need to incorporating tides, currents, and meteorological forecasts into their planning, recognizing that these environmental factors significantly impact search patterns, vessel maneuverability, and fuel consumption. Adverse weather conditions, such as the thick ice and darkness in this specific example, can also affect visibility and response times. Hence, at more advanced, professional levels of competence, voyage planning becomes a dynamic, multi-layered process that involves risk assessment, contingency planning, and collaborative decision-making. The aim is to ensure that students move beyond the assumption that a planned route will unfold exactly as expected and instead learn to anticipate potential disruptions.

3 From Showing to Involving: Rethinking Demonstration in Briefings

In the context of simulation-based training and pedagogy, demonstrations can be defined as the intentional and structured presentation of a skill, procedure, or system which is aimed at illustrating correct performance or operational principles (e.g., Sellberg & Lundin, 2017). Demonstration is typically incorporated as a component of the briefing process in maritime simulation training. As previously noted, the planning, execution, and monitoring of simulations vary considerably depending on the level of complexity. While a basic communication exercise may focus primarily on discrete tasks, such as the transmission of a simple verbal message via radio, these tasks constitute only a small part of the activity in full mission bridge simulations, where multiple tasks are carried out simultaneously. Nevertheless, across this range of complexity, briefings commonly include some form of demonstration. This may involve a basic walkthrough of equipment functionality, the correct handling of tools such as fire nozzles, or a procedural overview of complex operations, such as port arrival or the configuration of a vessel for dynamic positioning. Despite its routine inclusion in simulation training, demonstration poses specific pedagogical challenges. Although limited research has been conducted in this area, findings by Crouch et al. (2004) suggest that demonstrations that do not actively engage students result in no greater conceptual understanding than when no demonstration is provided. This observation has been corroborated in simulator instruction at Chalmers University of Technology, where it is frequently noted that students fail to apply demonstrated skills or knowledge effectively in simulation scenarios when the demonstration has relied on passive observation, lacked motivational framing, or did not invite participation. In this context, various strategies have been employed to address these challenges and adapt demonstration to different instructional contexts. In the case of relatively simple procedures, such as equipment setup, engagement can be increased by inviting students to participate through guided questioning, prompting them to propose each step and predict expected outcomes. Alternatively, an approach that allows students to work independently with detailed task instructions, manuals, and access to simulated or physical equipment has shown promise. This method encourages individual problem-solving and is typically followed by student-led demonstrations and feedback in later sessions.

For more complex simulation scenarios, a *flipped classroom model* has proven beneficial for our students (see e.g., Akçayır & Akçayır, 2018). Within this approach, students are expected to engage in structured preparatory work prior to the simulation session, including self-study of relevant procedures, group-based planning activities, and discussion of the upcoming task. This preparatory phase enables students to develop initial understandings and strategies that they can bring into the simulation briefing. Rather than entering the session with no frame of reference, students arrive with preliminary conceptions of how the task might be performed, conceptions that can then be refined, challenged, or adjusted during the instructor-led demonstration or through peer dialogue. In this way, the briefing becomes more than a demonstration of skills to passively watch; it becomes a dialogic and exploratory phase where students test their assumptions, clarify uncertainties, and make more meaningful connections to the simulation objectives. The demonstration, therefore, shifts from a unidirectional presentation to a collaborative process of sense-making. This engagement enhances both the relevance and the retention of the material being demonstrated, as students are more likely to recognize its applicability and reflect on its implications for their own planned actions.

However, in order for students to arrive prepared for such briefings, their engagement with the learning objectives must be carefully supported. This includes communicated expectations, access to appropriate preparatory materials, and an understanding that the learning process begins well before the students enter the simulator. When these conditions are met, we have seen how the flipped classroom approach not only increases student autonomy and accountability but also deepens the pedagogical value of both the briefing and the simulation that follows.

4 Setting up a Space for Briefing

Designing appropriate course materials for a simulation briefing requires careful attention to how the learning objectives and scenario are communicated to students. Learning objectives should be clearly articulated and designed to guide the participants' focus and actions during the activity, from briefing, through scenario, to debriefing (Sellberg, 2018). Hence, the learning objectives provide a foundational framework that shapes the content and structure of the briefing materials. How they should be presented, however, is a less straightforward question. For example, a PowerPoint presentation is commonly used, and can be an effective tool, offering a sequential format that unfolds slide by slide, directing attention towards key elements of the upcoming scenario. Hence, a PowerPoint might serve as a good foundation when preparing for briefing, but there are several other formats that can be considered.

In contrast to a pre-designed PowerPoint, the use of whiteboards or blackboards allows for dynamic interaction, enabling participants to draw, annotate, or modify visuals in real time (Fig. 2). For challenges such as navigational problems, which often involve abstract reasoning about spatial relationships or dynamic systems, drawings facilitate the externalization of these abstract ideas. I this way, drawings support collaborative discussions by providing a means for the instructor to demonstrate how an experienced professional reasons through navigational challenges while also creating an arena for participants to contribute ideas and address uncertainties. Moreover, unlike static materials, whiteboards and blackboards enable iterative refinement of navigational problems and their solutions, as participants can erase, adjust, or build upon existing drawings. This is a critical aspect of professional practice, where the primary challenge lies in understanding and responding to complex, ill-structured problems for which solutions are not readily apparent, and problem identification and problem solving evolve as new information and reasonings emerge through the collaborative efforts of a team (Schön, 2017).

While the whiteboard/blackboard format can be used to make such abstract problems more concrete for students, using physical or digital maps in briefing also

Passing Great Belt

Show your intentions!
Turn with autopilot r0,5M or handsteering
ARPA vs. TRAILs vs. visual conduct?
Safe speed 16 kt or 12 kt
Radar Range?
Crossing ferries follow COLREG!!!
Using PIs on good radar target

Fig. 2 An instructor using the whiteboard to demonstrate differences between propulsion systems

provide concrete and indexical references to the task at hand. As rather simple representations, maps afford several pedagogical advantages (Sellberg & Lundin, 2017). For students, maps afford a familiar representation of the marine environment and are one of the key working materials for visualize and plan their upcoming task. They provide a shared frame of reference for spatial orientation, allowing participants to establish a mutual understanding of the task and discuss the spatial dynamics of the scenario collaboratively. This shared understanding is particularly critical in joint route planning, where participants can engage in detailed discussions about the specific details of the environment. In these ways, maps provide an important recourse for spatial orientation and joint route planning in briefings, making the upcoming task concretely grounded in the specific details of the scenario. By grounding the briefing in the specifics of the scenario, maps make otherwise abstract planning tasks contextually relevant and thus actionable. When using maps projected on top of whiteboards or blackboards, maps can enhance collaborative problem-solving by integrating representational flexibility of drawings with the task-specific precision of the map. In particular, while the whiteboard can be used to model professional reasoning or explain navigational challenges in a flexible way, the map can anchor these discussions in the concrete, spatial realities of the upcoming scenario. The use of drawings and maps not only fosters a richer understanding of the task but also mirrors the integration of abstract reasoning and practical tools that professionals rely on in their work practices.

Moreover, three-dimensional representations, such as model ships, offer a tangible and spatial way to illustrate maneuvering problems. In our empirical data of maritime pilots in training, the use of model ships in conjunction with a representation of a waterway or port environment allows participants to explore the dynamics of navigation in a controlled and visually accessible manner. These models can be manipulated to demonstrate the effects of various factors, such as turning radii, wind, currents, or the interaction between vessels in close proximity (Fig. 3). For example, a model ship can be positioned to show how different rudder angles or engine configurations influence the vessel's movement, helping participants understand the principles of ship handling. Additionally, such models can illustrate spatial constraints, such as narrow channels or berthing challenges if placed on top of maps or drawings on the whiteboard/blackboard, enabling learners to visualize potential hazards and experiment with strategies for avoiding them already before entering the simulator. By engaging directly with the physical models, participants can better grasp abstract concepts through concrete tactile and visual means, facilitating professional reasoning and practical problem-solving in professionally relevant situations.

The use of video can serve as concrete illustrations of maritime scenarios, showing, for example, how accidents occur at sea. Videos provide a dynamic way to represent movement, environmental conditions, past incidents and their outcomes, offering rich contextual details that other formats often lack (Sablić et al., 2021). In the context of nursing education, videos that show how different clinical procedures are performed by an expert nurse have been tested as part of pre-briefing design (Coram, 2016). The videos used in this study closely mirror the scenarios assigned

Fig. 3 An instructor using ship models to demonstrate maneuvering dynamics, and drawings on the whiteboard to highlight movements

to students, with all actions and interactions carefully scripted to align with the educational objectives produced to match the scenarios later assigned to students, showcasing expert acute-care nursing practices. Experienced male and female nurses were recruited to demonstrate professionally exemplary behavior. Patients, on the other hand, were portrayed by a male and a female actor with backgrounds in theater and drama. All actors followed pre-approved scripts to ensure realism and professionalism of scenarios. The study showed that using these videos had a significant positive effect on students' clinical judgment during simulation, which led to new pre-debriefing routines for nursing students (Coram, 2016). While this can be a useful approach also for maritime pre-briefings, there is much to be learnt from mistakes, errors, and accidents. For instance, a video of an accident, e.g., the *Milano Bridge* incident in Busan 2020, can offer a detailed visual account of the ship's excessive speed and its subsequent failure to slow down as it attempted to berth. By capturing the vessel's movement in relation to the dock and the gantry crane, the video can be used to highlight and explain how the ship's speed, trajectory, and angle of approach contributed to the collision. Additionally, slow-motion replays or freeze frames can emphasize critical moments, such as when it became evident that the ship was not decelerating adequately or when evasive actions were initiated too late. When paired with verbal explanations, videos can serve as an effective teaching tool, drawing attention to key lessons, such as the importance of controlling the ship's speed, adjusting the rudder angle appropriately, and maintaining awareness of environmental conditions like wind, currents and port structures. Hence, videos can show the meaning of general principles of good seamanship in the concrete and specific circumstances of a real, practical example.

The spatial, material, and social arrangements of the room also structure the briefing session in specific ways, shaping interaction and participation while aligning with the learning objectives. For example, in Sellberg (2018), the briefing is held in a classroom adjacent to the full mission simulators. In the briefing room, the desks are arranged in a horseshoe shape, with designated places for each bridge team (Fig. 4). This layout fosters collaboration and establishes the focus on teamwork even before the scenario begins. In our study on desktop simulation, students are seated individually in rows in front of their computers, with the instructor positioned at the front of the class (Gyldensten et al., 2023). This arrangement supports a different mode of engagement, emphasizing individual work over collaboration, encouraging students' personal focus on the simulation tasks. In both contexts, the instructor occupies a traditional position at the front of the room, using a whiteboard and a sequentially organized PowerPoint presentation to guide the session in a structured manner. The physical layout aligns with an instructor-centered approach in MET, displaying the instructors' important role in supporting students' professional development (Sellberg, 2018). In different situations across my different data corpuses, part of the briefing takes place in the full mission simulator. In such instances, participants gather around the chart table, checklists, instruments on the bridge or orient towards the marine environment as seen through the windows. Here, the spatial and material configuration of the simulator is closely integrated with the professional setting, embedding the briefing within the specific context of the task. This arrangement supports instructions in regards to the professional demands of the scenario, allowing the instructor to show the meaning of general advice in the specifics of the upcoming scenario (Sellberg & Sharma, 2024).

In sum, the prepared teaching materials should be designed to establish joint attention on specific aspects of the activity, ensuring that critical points are elaborated upon in sufficient detail, while secondary elements are briefly addressed keeping focus on the specific challenges of the task. Setting ground rules for simulation, and communicate them to students, help set realistic expectations on what can be learned through simulations and facilitates a safe space for collaborative learning. It

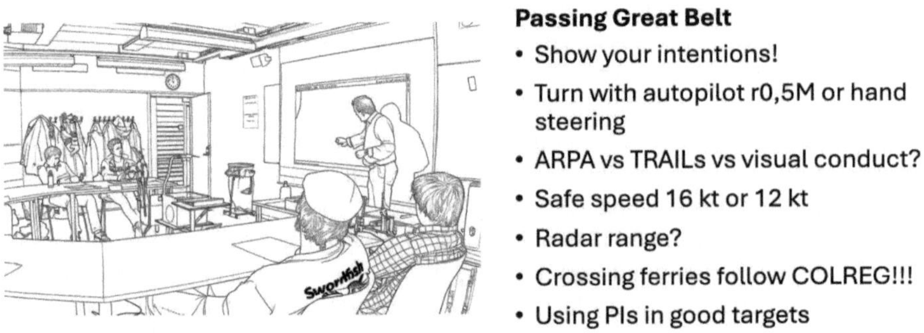

Fig. 4 Using a PowerPoint in the briefing to deliver prospective instructions

is also worthwhile to reflect on the how the spatial layout of the room shapes the organization and dynamics of the briefing session, and how the layout supports the intended learning outcomes of the simulation. Moreover, it is important to keep in mind that the different social, material and spatial arrangements are closely tied to the size of the groups, the intended learning outcomes and the recourses at hand at different simulator centers. Smaller groups and collaborative objectives often call for more interactive and shared spaces, while larger groups or individually focused tasks benefit from configurations that prioritize autonomy and minimize cross-talk between students.

5 The Briefing Phase

Previous MET research of briefing in practice has shown how and why the briefing is far from a straightforward delivery of instructions with clear outcomes; instead, it is a complex educational activity that needs to be carefully designed to lay the groundwork for the subsequent scenario (Sellberg, 2018). In particular, the challenge for the students lies in translating the prospective and often general instructions from the briefing into correct actions during the scenario (Sellberg, 2018). For the instructors, their challenges pose a practical problem: to deliver instructions that are open enough to cover a wide range of possible events that may or may not occur during the scenario, while being specific enough to provide meaningful and actionable information for students. The briefing thus becomes a crucial phase, not only for introducing the day's task but for supporting students in interpreting and acting upon open-ended instructions within the dynamic conditions of maritime navigation (Sellberg, 2018). In this way, the briefing phase underscores the complex relationship between instruction and action, highlighting its foundational role in shaping professional competence.

6 Delivering Instructions in the Briefing Phase

The content of the briefing typically weaves together general learning objectives with details specific to the scenario. For instance, when the task involves navigating the Great Belt, the instructor uses the context to discuss the application of anti-collision rules (COLREG) and the appropriate use of navigational instruments. Directives such as "Show your intentions!" or "Crossing ferries follow COLREG!!!" are presented as *maxims*, providing guidelines that frame the students' upcoming actions without dictating specific steps (Fig. 4). The use of maxims in instructional settings function as general principles that guide practice. However, in order for them to be useful, they require nuanced adaptation to the specific circumstances that might unfold during activities. In a study on sports coaching, Evans and Lindwall (2020) illustrate how a coaching intervention unfolds through the progressive unpacking of a maxim, starting with its initial formulation, followed by embodied enactments that demonstrate its application, and culminating in an exploration of its limitations and necessary adjustments. Rather than treating the maxim as a rigid rule, the instructional demonstration made by the basketball coach actively works through cases where its direct application might lead to unintended consequences. By contrasting correct and incorrect applications, the players are encouraged to develop a more flexible, context-sensitive understanding of the general principle presented as a maxim. The study highlights that in fields such as sports, surgery, or, as in our case, maritime navigation, proficiency depends on the ability to handle unexpected development in the course of events. Hence, maxims serve as foundational principles rather than fixed prescriptions, and their value lies not in their clarity alone but in their capacity to be shaped, tested, and refined through practical demonstrations of different actions. The open-ended nature of these maxims serves three key purposes. First, it accommodates the variability inherent in the scenario, as each bridge team starts from a different position and navigates to a unique destination. Second, students are unaware of the specific challenges they will encounter, such as traffic density or visibility, making the instructions necessarily preparatory rather than prescriptive, Third, these maxims encapsulate key navigational principles while allowing for the judgment necessary to respond to shifting conditions. The emphasis placed on these maxims in briefings, often reinforced through repetition and marked urgency, underscores their importance not only within the simulated environment but also in preparing students for their future roles, where their ability to interpret and apply COLREGS and principles of good seamanship in real time is critical for safe navigation.

In addition to maxims, the briefings we have analyzed include questions such as "Radar Range?" or "ARPA vs TRAILS vs visual conduct?" While these may appear open-ended and directed at the students, they often function as prompts for the instructor to elaborate on the situational use of bridge instruments. These elaborations serve to model expert reasoning and reinforce the expectation that students remain attentive to the specific circumstances of the scenario, adapting their actions accordingly. However, when we look more closely at the interactional design of these questions, and consider findings from research on classroom discourse, they closely resemble what has been described as questions with known answers or display questions (Macbeth, 2003). Such questions are asked not to seek new information but to elicit a specific, known response from the student, often as a means of initiating a teaching moment. Classic examples, like "What time is it, Denise?" (Mehan, 1979), illustrate how these questions operate within the well-documented Initiation–Response–Evaluation (IRE) sequence, where students are positioned as respondents whose contributions are evaluated based on correctness rather than interpretive reasoning.

In our data, simulation briefing questions frequently follow this pattern. Rather than opening up dialogue, they function as cues for the instructor to guide attention to relevant instruments or situational features, often followed by explanatory talk. While this serves important pedagogical functions, focusing student attention, demonstrating expert judgment, and reinforcing professional accountability, the question design also constrains the interactional space available to students. As Macbeth (2003) and others have shown, such questioning formats tend to prefigure a narrow range of acceptable responses, limiting opportunities for students to publicly engage in professional reasoning or to articulate their own interpretations of the scenario.

If the instructional goal is to support students in developing professional knowledge and accountable reasoning practices, then the strategic use of genuine questions becomes crucial. Genuine questions, that is, questions for which the instructor does not already know the answer and where multiple interpretations are possible, have been shown to foster more dialogic and participatory forms of engagement (Roth, 1996). Unlike questions with known answers, genuine questions create interactional space for students to publicly reason through their observations, justify their decisions, and engage in collaborative sense-making with instructors and peers. In this way, they not only assess student understanding but also contribute directly to the development of that understanding in practice. In the context of our data, this distinction becomes particularly clear. Instead of asking "*Radar Range?*" one might instead ask, "*Given that we are in confined waters and considering the traffic density, what radar range do you think is most appropriate, and why?*" or "*How would you use ARPA functions to monitor potential collision risks in this situation?*" These are genuine questions in that they invite reflection, interpretation, and reasoning grounded in the particulars of the scenario. They prompt students to consider the affordances and limitations of radar and ARPA in context and to articulate a rationale for their choices. By designing questions this way, instructors shift the interactional footing of the briefing. Rather than positioning students as recipients of expert knowledge, they are treated as accountable participants who must navigate uncertainty, weigh alternatives, and justify their planned use of navigational tools. This fosters a more participatory and reflective learning environment, one that mirrors the situated nature of professional judgment and decision-making in real-world maritime operations.

7 Facilitating Discussions in the Briefing Phase

As demonstrated in our analysis of questions, question design is central to facilitating discussions in the briefing phase, inviting students to think beyond the immediate task and engage in joint professional reasoning. Fostering this capacity goes beyond technical instruction; it also involves creating opportunities for students to reason, explain, and justify their planning in dialogue with others (Littleton & Mercer, 2013). Drawing on Neil Mercer's concept of *exploratory talk*, a dialogic approach to teaching and learning emphasizes the importance of open, collaborative dialogue in which ideas are shared, questioned, and developed collectively (Mercer, 2019). In contrast to the IRE format that is common in classroom interactions, exploratory talk allows students to make their thinking visible, consider alternative viewpoints, and collaboratively construct understanding. Carefully designed questions play a key role in enabling this kind of dialogue, inviting students not only to demonstrate what they know but to actively participate in the construction of navigational knowledge.

To facilitate exploratory talk, Mercer (2019) highlights the importance of open, genuine, and collaborative question designs. The types of questions most conducive to this kind of dialogue include:

- Open-ended questions invite a range of responses and prompt students to explain their thinking. Example questions: "What do you think is the most important consideration when setting up the radar in this area?" or "How would you approach planning this route through coastal waters?"
- Speculative or hypothetical questions are useful for encouraging reasoning through alternatives or thinking ahead. Example questions: "What if visibility suddenly decreases, how would that change your use of ARPA?" or "Suppose another vessel starts changing course, how would you assess the situation using radar and visual observations in this situation?"
- Why and how questions encourage students to unpack their decisions and reasoning. Example questions: "Why did you choose that particular radar range for this leg of the passage?" or "How could your choice of CPA settings affect your monitoring of nearby vessels?"
- Follow-up or probing questions deepen understanding and extend contributions. Example questions: "How did you assess the risk of close-quarters situations in your plan?" or "What other information should you keep track of in these waters?"
- Collaborative framing questions invite students to reason together and reflect on each other's input. Example questions: "Do you agree with that route plan? Why or why not?" or "How could we combine your observations with what was just said to improve our situational awareness?"

These types of questions not only engage students in doing navigation but also in thinking about what counts as good professional practice and why.

8 Discussion

In this chapter, we have emphasized the crucial role of briefings in simulation-based learning as a means of situating theoretical knowledge within the context of real-world practice. While we know that time constraints are an inevitable challenge in simulation-based training, we still want to stress that ensuring adequate time for briefing is essential to making the most of simulator time. Briefings are not just preparatory but foundational to a focused and effective learning experience. By setting clear expectations, introducing specific learning objectives, and providing context for the upcoming scenario, briefings help students bridge the gap between abstract concepts and the practical challenges they will encounter. However, to be truly effective, briefings must extend beyond the mere transmission of information.

Active involvement is key to this process. Rather than simply "showing" and "telling", demonstrations should invite learners to engage with the material through participation, prediction, and reasoning. Encouraging self-directed study and collaborative planning before the briefing enriches the session, allowing for critical discussion, clarification, and correction that deepens the learners' understanding. Rushing through this critical phase can undermine its value, leaving students unprepared to handle the complexities of the simulation. By dedicating sufficient time to briefings, instructors create a space for students to understand the task at hand, reflect on their roles, and develop strategies for the scenario. This investment in time ensures that the subsequent simulator activities are purposeful, aligned with learning goals, and maximally effective in fostering both skill development and professional judgment.

Finally, psychological safety should be understood not as the absence of discomfort, but rather as the ability to engage with uncertainty under fair and transparent conditions. In high-stakes professions, particularly those involving leadership, students must learn to manage ambiguity and uphold accountability. This requires clear ground rules, collaborative norms, and a structured orientation to roles, systems, and contextual conditions, ensuring that students are supported in their professional development. This approach fosters an environment where students can engage actively and reflectively, contributing to fostering future mariners who uphold the highest standards of practice and professional integrity.

References

Akçayır, G., & Akçayır, M. (2018). The flipped classroom: A review of its advantages and challenges. *Computers & Education, 126*, 334–345.

Coram, C. (2016). Expert role modeling effect on novice nursing students' clinical judgment. *Clinical Simulation in Nursing, 12*(9), 385–391.

Crouch, C., Fagen, A. P., Callan, J. P., & Mazur, E. (2004). Classroom demonstrations: Learning tools or entertainment? *American Journal of Physics, 72*(6), 835–838.

El Hussein, M., Harvey, G., & Kilfoil, L. (2021). Pre-brief in simulation-based experiences: A scoping review of the literature. *Clinical Simulation in Nursing, 61*, 86–95.

Evans, B., & Lindwall, O. (2020). Show them or involve them? Two organizations of embodied instruction. *Research on Language and Social Interaction, 53*(2), 223–246.

Gyldensten, W., Sellberg, C. & Wiig, A. C. (2023). Monitoring, noticings and instructional feedback during maritime desktop simulation. *Copenhagen Multimodal Day*, Copenhagen, Denmark, October 6, 2023.

Kolbe, M., Grande, B., & Spahn, D. R. (2015). Briefing and debriefing during simulation-based training and beyond: Content, structure, attitude and setting. *Best Practice & Research Clinical Anaesthesiology, 29*(1), 87–96.

Littleton, K., & Mercer, N. (2013). *Interthinking: Putting talk to work*. Routledge.

Macbeth, D. (2003). Hugh Mehan's Learning Lessons reconsidered: On the differences between the naturalistic and critical analysis of classroom discourse. *American Educational Research Journal, 40*(1), 239–280.

McDermott, D. S., Ludlow, J., Horsley, E., & Meakim, C. (2021). Healthcare simulation standards of best practice™ prebriefing: Preparation and briefing. *Clinical Simulation in Nursing, 58*, 9–13.
Mehan, H. (1979). *Learning lessons: Social organization in the classroom.* Harvard University Press.
Mercer, N. (2019). *Language and the joint creation of knowledge: The selected works of Neil Mercer.* Routledge.
Roth, W. M. (1996). Teacher questioning in an open-inquiry learning environment: Interactions of context, content, and student responses. *Journal of Research in Science Teaching: The Official Journal of the National Association for Research in Science Teaching, 33*(7), 709–736.
Sablić, M., Mirosavljević, A., & Škugor, A. (2021). Video-based learning (VBL)—past, present and future: An overview of the research published from 2008 to 2019. *Technology, Knowledge and Learning, 26*(4), 1061–1077.
Schön, D. A. (2017). *The reflective practitioner: How professionals think in action.* Routledge.
Sellberg, C. (2018). From briefing, through scenario, to debriefing: The maritime instructor's work during simulator-based training. *Cognition, Technology & Work, 20*(1), 49–62.
Sellberg, C., & Lundin, M. (2017). Demonstrating professional intersubjectivity: The instructor's work in simulator-based learning environments. *Learning, Culture and Social Interaction, 13*, 60–74.
Sellberg, C., & Sharma, A. (2024). Toward multimodal learning analytics in simulation-based collaborative learning: A design ethnography of maritime training. *International Journal of Computer-Supported Collaborative Learning*, 1–21.
Sharma, H., Patil, A. D., & Baviskar, A. (2023). Fiction contract: Its importance in simulation-based medical education. *International Journal of Basic & Clinical Pharmacology, 12*(5), 766.
Somerville, S. G., Harrison, N. M., & Lewis, S. A. (2023). Twelve tips for the pre-brief to promote psychological safety insimulation-based education. *Medical Teacher, 45*(12), 1349–1356.
Starup, M., Sellberg, C., & Wiig, A. C. (2024). Playing to learn? Analyzing participants' framing of competition and professional conduct in maritime simulations. *Learning, Culture and Social Interaction, 46*, 100821.

Charlott Sellberg is an Associate Professor in IT and Learning at the Department of Applied IT, University of Gothenburg. With a multidisciplinary background in Cognitive Science and Human-Computer Interaction, she earned a PhD in Education from the University of Gothenburg in 2018. Her doctoral work examined both the theoretical and practical aspects of simulation-based training in the maritime industry. Since then, she has continued to study the role of simulations in maritime training and assessment, in areas such as basic safety training, SAR operations, and advanced ship handling, continuing to deepen the understanding of simulation as a learning environment and informing educational practices within the maritime domain.

Daniel Ernstsson is a former deck officer with extensive experience in navigation and offshore operations within the maritime industry. He currently serves as an instructor in nautical science at the Division of Mechanics and Maritime Sciences at Chalmers University of Technology in Gothenburg. In his role, Daniel is primarily involved in the bachelor's program in nautical science and various professional training courses for the maritime industry. He has several years of experience as a course examiner and simulator instructor, working with a wide range of training formats—from desktop simulations and tabletop exercises to full-mission bridge simulations and maritime safety and survival training.

The Instructor's Work During Simulations

Susan Harrington

Abstract This chapter outlines a comprehensive guide to the creation of a fruitful learning environment for simulation exercises. In this respect, the role and importance of monitoring students' ongoing activities during simulations is highlighted as the basis of being able to support the students in their learning processes. In addition, the chapter outlines how role-plays in the simulated environment can function as an arena for professional learning, using the threshold concept framework to illustrate the messy, non-linear process of becoming a seafarer.

Keywords Simulation-based training · Threshold concepts · Instructor roles · Professional identity formation · Maritime education and training (MET)

1 Introduction

In the maritime field, and others where simulation is used in professional learning, the instructor's role is pivotal. Simulation provides the opportunity for experiential learning in a controlled environment, where students can safely learn from mistakes under the guidance of experienced instructors, thus developing the competences required for successful maritime operations. In order to train students who are prepared for real-world challenges, the instructor's role is crucial before, during, and after simulation, in both setting educational objectives and supporting students in achieving them (see Table 1). Briefly, instructors are responsible for designing and facilitating simulated scenarios that not only replicate the professional maritime

S. Harrington (✉)
Department of Applied Information Technology, University of Gothenburg, Gothenburg, Sweden

Department of Computer and Information Science, Linköping University, Linköping, Sweden
e-mail: susan.harrington@ait.gu.se; susan.harrington@liu.se

C. Sellberg (ed.), *Simulations in Maritime Education and Training*, WMU Studies in Maritime Affairs 12, https://doi.org/10.1007/978-3-032-08394-4_4

Table 1 The instructor's role in simulation-based training

Scenario design and preparation	Instructors are tasked with designing realistic training scenarios that reflect the challenges and situations that maritime professionals may encounter at sea. This involves crafting appropriate simulation exercises that align with the learning objectives and the specific competencies that students need to develop. The scenarios may include emergency situations, navigation challenges, and operational procedures relevant to the students' future roles.
Facilitation of simulator training sessions	During the simulation exercises, instructors guide students through the scenarios, ensuring that they understand the objectives and the operational procedures involved. They provide clear instructions and support, helping students to engage fully with the simulation. Instructors must also monitor the students' performance, ensuring that they adhere to safety protocols and best practices as appropriate.
Real-time assessment and feedback	One of the most critical aspects of an instructor's role is providing real-time assessment and feedback during the simulation. Instructors observe students' decision-making processes and actions, offering constructive feedback that helps them to reflect on their performance. This feedback is essential for reinforcing learning and correcting any misconceptions or errors in judgment. Instructors may use formative assessment techniques to evaluate students' competencies and provide guidance for improvement.
Debriefing and reflection	After the simulation exercise, instructors conduct debriefing sessions where they work with students to reflect upon performance and discuss key learning points. This reflective practice is crucial for consolidating knowledge and enhancing understanding. Instructors facilitate discussions that encourage students to articulate their thought processes, evaluate their decisions, and learn from both successes and mistakes. The debriefing process is an opportunity for instructors to reinforce important concepts and ensure that students can apply their learning in future scenarios.
Adaptation and continuous improvement	Instructors must remain adaptable and responsive to the needs of their students. They may need to modify scenarios on-the-fly based on the students' performance and specific learning needs. Additionally, instructors are responsible for continuously improving their own teaching methods and the training program based on factors such as feedback from students and advancements in maritime education practices.
Integration of industry standards	Instructors ensure that the training aligns with industry standards and regulatory requirements, such as those set by the International Maritime Organization (IMO) and the Standards of Training, Certification, and Watchkeeping (STCW). This alignment is crucial for preparing students to meet the expectations of the maritime industry and to enhance safety at sea.

environment but also align with specific learning objectives (see chapter "Designing Simulation for Learning" for more on scenario design). The role of instructors during pre-briefing (see chapter "Preparing for Simulation Through Briefing") and debriefing (see chapter "Facilitating Debriefings for Feedback, Discussion, and Reflection") are discussed in the relevant chapters of this book—here we go into more depth as we explore the crucial role played by instructors as the simulation is running.

Guiding students through simulations involves an ongoing cycle of formative assessment, where instructors monitor student performance and provide feedback—directly and indirectly—which is intended to prompt critical thinking and self-reflection amongst the students (Karahalil et al., 2023). As we progress through this chapter, we discuss the instructor's work during simulation as we model the cycle of monitoring, assessment, and feedback (Sellberg, 2018; Sellberg & Sharma, 2024), and three general mechanisms for professional learning from simulations (Sellberg & Solberg, 2024). We also focus on the function of role-plays in the simulated environment, and how these contribute to students' development and mastery of professional competences.

2 The Instructor's Work During Simulation-Based Training.

Figure 1 illustrates the ongoing cycle of monitoring, assessment, and feedback sustained by instructors during simulation. The instructor's role is one of support in ensuring that the process of co-creation of knowledge, involving the instructor, the students, and their interactions with the various instruments in the simulator environment, progresses towards the intended learning outcomes (Sellberg, 2018). The instructor's role here is one of facilitating the students in their journey towards becoming seafarers, using the three mechanisms identified by Sellberg and Solberg (2024) as key to translating simulated events into professional practice (see Fig. 2).

The first mechanism involves the instructional practice of "filling in" aspects of the work context that might be missing in the simulator. This relates to the concept of fidelity in simulation, or how closely the physical simulator environment and the events comprising the simulated scenario resemble reality. Although the term 'fidelity' appears often in literature on simulator-based training, there is no commonly accepted definition of exactly what the term refers to. Depending on the context, 'fidelity' may refer to how closely the physical bridge and the instruments it

Fig. 1 Instructor's role during simulation

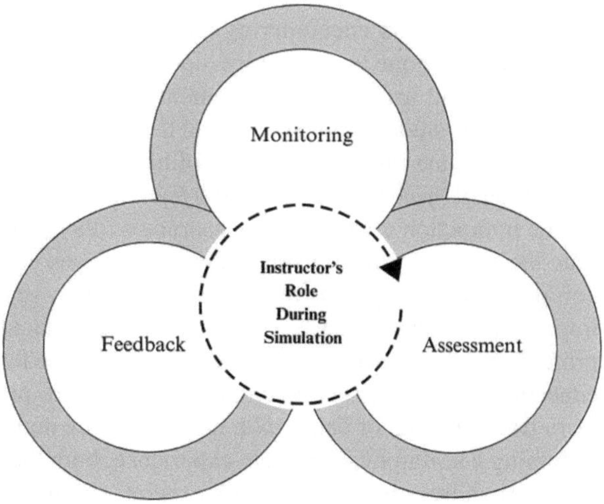

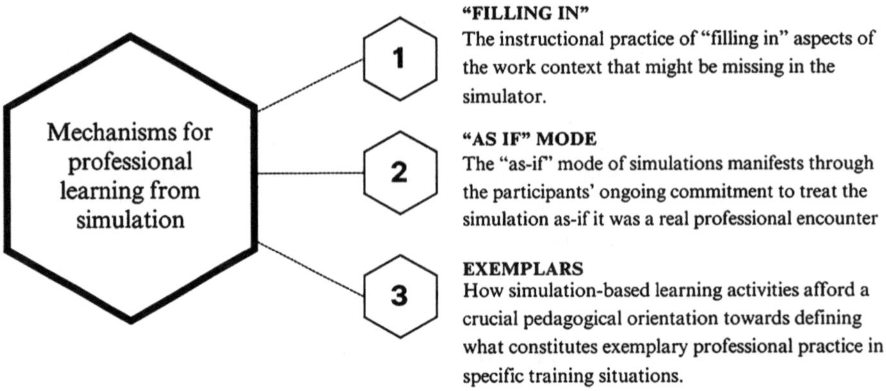

Fig. 2 Mechanisms for professional learning from simulations (adapted from Sellberg & Solberg, 2024)

contains match that found on a particular vessel, sociotechnical elements of the simulation, or visual and auditory cues occurring during the exercise (Sellberg & Solberg, 2024). Harrington et al. (2025) propose the term "*functional congruence*" as being more appropriate here, defining it as "*where simulated scenarios align meaningfully with real-world skills needed to meet learning objectives*". The instructor's role here is to bridge the gap between the simulator environment and the real work context, creating as realistic a learning environment as possible, thus enabling students to become immersed in their simulated roles, experiencing the complexity of their future professional role, and progressing along their journey of becoming their future professional self.

Sellberg (2017) describes how instructors use talk and gestures in this practice of filling-in missing elements of the work context, aiding students in their understanding of the vessel their simulation is situated in, as they develop ship sense. In discussing a situation where second-year master mariner students are required to take evasive action upon encountering another vessel, the instructor's movement from the instructor room to the bridge upon seeing that the students required further instruction, and use of hand movements and gestures to illustrate his instructions and guide the students towards taking the correct course of action are described. In this instance, the instructor's tilting of his hands represents the rudder in his verbal instruction of "*you'll start with just a few degrees rudder*" (Sellberg, 2017, p. 2321). As the interaction continues, the instructor tilts his head from left to right to illustrate the sway of vessel, using facial expressions and hand gestures to emphasise key points in his spoken instructions. This non-verbal interaction represents an important element in the learning process, as students learn from the instructors' professional experience of the work context. While the simulator can be a powerful educational tool, there is much about their future profession that students cannot learn from reading, or being told, how things are done. Much of what is involved in becoming a seafarer comes from experience, both the students' own and that which is passed on from interactions with experts—in this case, the instructor.

The second mechanism—sometimes labelled the "as-if" mode of simulations—manifests through the participants' ongoing commitment to treat the simulation as-if it was a real professional encounter. As the value of simulator-based training comes from the immersive nature of the experiential nature it offers, it is essential for students to behave as though they are in their assigned role, encountering the events in the simulation on a real vessel where their actions and decisions have consequences. Engaging in these roles and overlooking the ways in which the simulation does not align with reality, thus suspending their disbelief. Everyone involved in the simulation knows that, for example, the physical simulator does not perfectly match the setting on board a vessel, and that the student assuming the role of captain does not really have authority over others but has collectively agreed to engage as though the scenario was real. Muckler (2017) outlines the importance of the ability to suspend disbelief in order to be properly immersed in the learning environment, and factors which can impact upon students' ability to do so, e.g., fidelity of the simulator, and the fiction contract.

Fiction contract	An implicit agreement established between instructors and students, aimed at fostering an environment conducive to learning. This contract acknowledges the simulated nature of the experience, while promoting a willingness amongst participants to engage with the scenario as if it were real. This fiction contract functions as a mutual understanding; it encourages learners to invest emotionally in the simulation while recognising its limitations. In doing so, it establishes a safe space for learning, where learners can practice skills and make decisions as though they were in authentic situations, yet within a controlled and supportive setting.

Students participating in simulation are responsible for their own willingness to invest themselves in the scenario, making decisions and practicing skills as they would in the real work context. At the same time, instructors have a responsibility to uphold their role in facilitating this suspension of disbelief. Part of this relates to the design of the training scenario, and choosing an appropriate simulator for the task at hand, but another major factor is the creation of a safe learning environment where students can feel comfortable in making mistakes. This psychological safety is essential for effective learning from simulation, and is largely constructed during briefing and debriefing sessions, and through providing students with meaningful feedback.

Psychological safety	The perception students have about the potential consequences of taking risks in the simulator. When students are confident that their mistakes will be handled in a respectful manner by the instructor and other students, it becomes easier to assume an unfamiliar role, accepting the potential of failure. Instructors must create an environment of trust, where the embarrassment of making a mistake in the simulator is not so great as to discourage students from immersing themselves in their roles (Muckler, 2017).

The third mechanism relates to how simulation-based learning activities afford a crucial pedagogical orientation towards defining what constitutes exemplary professional practice in specific training situations. In other words, what constitutes good seamanship? This is a question that is very difficult to provide a definitive answer to, due to the complexity of the professional context. Sellberg et al. (2021, p. 322) describe it as "*the kind of behaviour expected from an average good master or seaman in navigating and maintaining a vessel, usually associated with traits such as strong work ethics, good judgement, and a sense of responsibility*." However, each of the terms used here are very broad, and open to interpretation. Good seamanship will be different in different contexts, and assessing good seamanship in the simulator involves defining what behaviour would be expected in the specific scenario being simulated. The instructor's role in this third mechanism generally centres around fostering students into the norms of life as a professional mariner, although even this varies according to the type of training being undertaken.

According to one instructor participating in a study on core pedagogical concerns while engaging in the process of monitoring, assessment and feedback during simulation (see Fig. 1), good seamanship involves "*establishing positive norms as a means of this is what is expected of behaviour, conduct and so on and so forth*", while a colleague expands upon this by saying that "*It depends on the different type of simulations. In the BRM context we are definitely talking about fostering. We say that it's fostering a positive attitude towards the public... Fostering positive attitudes towards safety, good seamanship... So, in the BRM context, it's a lot about fostering, but in another course, we might not do that at all... If you look at the STCW it won't say to be fostered in a certain way. But if you read that between the lines it is about fostering.*" The instructor's role here is to "*read between the lines*" of the STCW, both in designing a scenario that trains a particular skill or competency, and in assessing how students demonstrate their proficiency.

Professional vision	"*Socially organized ways of seeing and understand events*" (Goodwin, 1994, p. 606). Professional vision develops as members of a profession develop a perspective based on the culture and norms of their community, and their own experiences. This allows experts (e.g., instructors) to notice and reason about events or information that may be overlooked by novices (e.g., students) or people outside of the profession. In terms of teaching, professional vision allows instructors to observe, monitor, and assess student performance through reflective practice (e.g., Wyss et al., 2021)

Instructors' professional vision and professional reasoning is at the core of the monitoring and assessment that occurs during simulation, and in constructing the feedback delivered to students before, during, and afterwards. Although instructors may not be able to define good seamanship verbally, their experience and expertise enables them to notice what is unfolding as the scenario progresses, and engage in "*knowledge-based reasoning*" (Gegenfurtner et al., 2020, p. 59) about what they have noticed. Returning to the study on instructor's core pedagogical concerns, this reasoning appears to be extended from the instructors to the students. Recounting how there may be more than one acceptable way to approach an event, one instructor explains how it is important to discuss the reasoning behind why students made particular decisions, rather than having a concrete exemplar performance that must be replicated by students—"*There is a certain aspect of priorities. So, I cannot tell a competent student sometimes that you didn't do this and therefore this is a fail. I need to ask the question, why do you have this setting? And if the answer is reasonable, thought through and with a proper kind of cause, then I cannot [fail the student].*" During the simulation, instructors must use their knowledge and experience to adapt to the unfolding scenario on the fly, guiding students towards behaviour that is considered to be good seamanship in the given context. Finally, as part of this third mechanism, instructors use feedback from students, and their own noticing around potential pain points in the scenario, to continually evolve and improve the scenarios used for simulator training.

3 The Function of Role-Playing in the Simulated Environment

Role-playing is a core element of simulator-based training, where students are encouraged to take on assigned professional roles and act accordingly during the simulated scenario, and instructors often assume the role of another vessel or authority. This offers a way in which students can link theory and practice, and for instructors to provide guidance and corrective feedback during a simulation. While simulation is used to train a plethora of skills and competencies, the overarching aim is to develop and refine the students' skills as future professionals. In Harrington et al. (2025), one student illustrates the immersion of the role-play as "*I become the captain […] I'm not afraid to give orders […] and it will be more* ***my*** *boat.*"

This role-playing offers a valuable opportunity for students to be socialized into the culture and norms of their chosen profession, connecting what they have learned and experienced so far in classes, cadetships, and previous work experience. The instructor plays several roles in this process—designing an effective scenario for simulation (see chapter "Designing Simulation for Learning" for more on scenario design), briefing students prior to the simulation (chapter "Preparing for Simulation Through Briefing"), and debriefing afterwards to help solidify learning (chapter "Facilitating debriefings for feedback, discussion, and reflection"). Alongside these, the instructor plays a pivotal role during the simulation itself, being the facilitator of a learning environment conducive to the type of deep, transformative learning required when students engage in such a process of 'becoming'. The instructor must tailor the simulation to the level of the students, monitoring progress as the scenario progresses, then interjecting or adjusting the exercise as appropriate to ensure that the intended learning objectives are achieved. Of course, this can be easier said than done, as students can approach events in an entirely different way than expected, taking actions and making decisions that instructors did not foresee when planning the simulation.

While instructors can plan according to the education and experience the students have had previously, it can be more complex to consider the messiness of the learning process. Learning is not a linear process, and can involve some back-and-forth over an extended period of time, as students master new skills and embed new knowledge. This is illustrated by Meyer & Land's (2003) threshold concept framework. This framework presents transformational learning as a three-stage liminal journey, where students start as novices in the pre-liminal phase, encounter new knowledge (culture and norms of becoming a seafarer) and embark on the liminal phase, before finally emerging into the post-liminal phase having become seafarers (see Fig. 3).

Threshold concept	A concept, mastery of which is "*akin to a portal, opening up a new and previously inaccessible way of thinking about something. It represents a transformed way of understanding, or interpreting, or viewing something without which the learner cannot progress. As a consequence of comprehending a threshold concept there may thus be a transformed internal view of subject matter, subject landscape, or even world view. This transformation may be sudden, or it may be protracted over a considerable period of time, with the transition to understanding proving troublesome. Such a transformed view or landscape may represent how people 'think' in a particular discipline, or how they perceive, apprehend, or experience particular phenomena within that discipline (or more generally)*" (Meyer & Land, 2003, p. 1)

In the university setting, simulations will be taking place in the students' liminal phase, meaning their exact level of knowledge when attending the simulator on a specific day, and therefore how they approach the scenario, can be difficult to pinpoint. Meyer & Land's (2003) threshold concept framework offers a useful tool for understanding the learning journey upon which the students are progressing, and the ongoing process of becoming. This process is often protracted, taking place over a number of years as the students engage in both formal education and practical experience during cadetships. In order to master a threshold concept, students must be emotionally invested in their learning, and willing to take on board the new knowledge and experiences they are encountering as they work towards their future professional self. The mastery of a threshold concept is what enables a learner to pass from being a student to being a professional, using the language of the profession, demonstrating the norms, and being accepted by others in the professional community as being one of their own. In other words, transforming from being a student

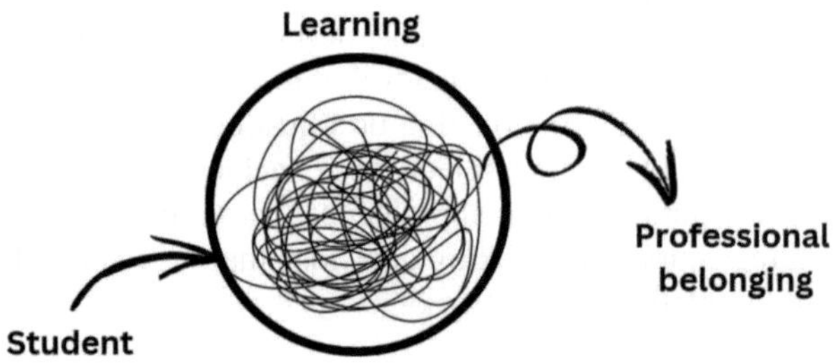

Fig. 3 Liminal journey from student to professional (adapted from Meyer & Land, 2003)

learning about seafaring, to being someone who can be trusted as a competent member of a team on board a vessel (see Table 2).

As yet, there is no research specifically focusing on threshold concepts within MET, however, Wilson-Thain & Hammond (2023) explore clinical reasoning in medical professionals using the threshold concept framework, elements of which may be transferrable. Clinical reasoning relates to the ability of a medical practitioner to accurately diagnose or form a hypothesis about the variety of symptoms a patient presents with and develop an appropriate treatment plan. This has similarities to the expectation of a seafarer being able to notice an unfolding event and proactively decide upon an appropriate course of action. Both involve complex, uncertain situations which have potentially catastrophic consequences if handed incorrectly. Training in high-fidelity simulators is one of the recommendations made by Wilson-Thain & Hammond (2023) as necessary for facilitating students'

Table 2 Characteristics of threshold concepts (adapted from Meyer & Land, 2003)

Characteristic	Definition
Transformative	Mastering a threshold concept changes the students' perspective, how they see their environment, and how they understand events happening around them
Integrative	Once the threshold concept has been mastered, the student can bring together different elements of their learning and experience, seeing and understanding the links between tasks, skills, and knowledge they have encountered during their training. In other words, students grasp and understand the interrelatedness of ideas within the profession.
Irreversible	Once a threshold concept has been mastered, it is impossible (or incredibly difficult) to unlearn. A student cannot return to being a novice once they have become a competent seafarer.
Troublesome	Due to the complex nature of threshold concepts, and the transformative aspect of mastering them, the new knowledge can often be difficult for students to take on board when first encountered. It may seem counterintuitive, conflict with their existing beliefs and previous knowledge, or challenge the assumptions they have made. This contributes to the difficulty of the liminal phase (or messy middle), where learning appears to go back and forth before evident progress is made.
Bounded	The threshold concepts within a discipline apply only to that discipline. Mastering the concepts necessary to become a seafarer, for example, relate only to the student's progress in that particular process of becoming, and do not necessarily increase expertise elsewhere.
Discursive	One of the defining characteristics of threshold concepts is the effect that their mastery has on the language used by the student. A student who has mastered the necessary threshold concepts begins to speak as a seafarer, using the language of their professional role, and can be seen to behave as a seafarer rather than a student. Their communication will reflect their deeper understanding.
Reconstitutive	Mastering a threshold concept tends to see the student begin to think as a professional. All of the skills, knowledge, and competencies they have encountered during their training have come together, equipping them with the repertoire needed to act in a way that represents good seamanship.

development of clinical reasoning, as is an integrative education and training program. The authors highlight the value of simulation in allowing students the opportunity to practice complex decision making in a safe environment, where students can fail without consequences. Kneebone (2009) explores simulation through the lens of threshold concept mastery, stressing the importance of learning tasks in context rather than as standalone skills to be memorized and repeated by rote, and the difference between what experts and novices consider to be "simple". The author also focuses on the uncertainty and messiness of the learning journey, and the importance of ensuring that simulation recreates these in so much as is possible. To return to the discussion on role-play during simulation, Kneebone (2009) discusses the need for students to take on different roles in the simulated scenario in order to challenge their assumptions and develop shared mental models around the reality of the clinical setting.

As such, the instructor's role while the simulation is in progress is to maintain an appropriately realistic learning environment for students to immerse themselves in their assumed professional roles. To this end, Wahl et al. (2025) discuss the importance of simulations as experiential learning where students can step into their future professional role, increasing their understanding of who they will be and what they will do. Through engaging in the process of monitoring, assessment and feedback shown in Fig. 1, the instructor is responsible for using their experience and expertise to guide students in developing their own professional self. Instructors design scenarios giving careful consideration to key pedagogical concerns and regulatory requirements, tailored to the students' level. Monitoring the simulation in progress involves utilizing professional vision and expert judgment to reason about student performance, tweaking scenarios and providing feedback as appropriate, and considering the students' own reasoning in making decisions. This supports the students in developing their ability to think and act as a seafarer, as they assume their future professional role. Finally, feedback is given in a way that supports the students' development, cementing the notion of the simulator as a safe learning environment where mistakes can be made and learned from.

4 Summary and Guidelines for Action

This final section provides a brief summary of the key points discussed in the chapter, alongside some guidelines for instructors to take action.

4.1 Summary of Key Points

Instructors' role: Instructors hold a central role in designing simulation scenarios, overseeing student engagement during simulations, and delivering constructive feedback. Their expertise is essential for guiding students in developing professional reasoning and competencies, and in maintaining a safe and effective learning environment.

Threshold concepts: Fundamental, transformative concepts that are often challenging to comprehend and are specific to individual disciplines. Their mastery facilitates significant changes in how students think, communicate, and act, equipping them for their professional roles.

Liminal phase: The process of mastering threshold concepts is characterised by a transitional period, often referred to as the "messy middle," during which students experience iterative progress as they refine their understanding. Instructors play a key role in ensuring that simulation is designed to support this process.

Role of simulation in professional learning: Simulations play a pivotal role in contextualised learning by supporting the development of professional behaviours and enabling students to assume their prospective roles. These simulations should replicate real-world complexities and uncertainties as closely as possible.

4.2 Guidelines for Action

Optimise scenario design: Design simulation scenarios that are functionally congruent with reality, contextually aligned with professional standards, and compliant with relevant regulatory and pedagogical requirements. Scenarios should be tailored to the students' level of expertise and learning needs (see chapter "Designing Simulation for Learning").

Encourage immersion: Foster a learning environment that facilitates the adoption of professional roles by students, encouraging them to challenge preconceived notions and develop shared mental models reflective of authentic professional practices.

Facilitate reflective learning: Implement assessment and monitoring tools to evaluate student performance during simulations, providing actionable feedback that deepens their understanding and supports improvement.

Maintain a safe learning environment: Emphasise the nature of simulations as risk-free spaces for experimentation and learning. Reinforce the notion that errors are integral to the learning process and provide constructive feedback to support growth.

Adopt an integrative approach: Ensure that simulation-based learning tasks are interconnected and delivered within a contextual framework, rather than as isolated skills or exercises in rote learning.

Promote role-playing: Encourage students to assume various roles within simulated scenarios to expand their perspective and cultivate a comprehensive understanding of professional practices.

The next chapter in this book outlines the facilitation of debriefings for feedback, discussion, and reflection.

References

Gegenfurtner, A., Lewalter, D., Lehtinen, E., Schmidt, M., & Gruber, H. (2020). Teacher expertise and professional vision: Examining knowledge-based reasoning of pre-service teachers, in-service teachers, and school principals. *Frontiers in Education, 5*. https://doi.org/10.3389/feduc.2020.00059

Goodwin, C. (1994). Professional vision. *American Anthropologist, 96*(3), 606–633.

Harrington, S., Klein, S., Fratoni, A. C., Sellberg, C., & Lindwall, O. (2025). Meaningful automated feedback? *Balancing tensions in participatory learning analytics design*. 18th international conference on computer-supported collaborative learning, Helsinki, Finland.

Harrington, S., Sellberg, C., & Lindwall, O. (2025). Crafting congruence: Towards systematic scenario design in maritime simulator training and assessment. *WMU Journal of Maritime Affairs*. https://doi.org/10.1007/s13437-024-00356-3

Karahalil, M., Lützhöft, M., & Scanlan, J. (2023). Formative assessment in maritime simulator-based higher education. *WMU Journal of Maritime Affairs, 22*(2), 181–207. https://doi.org/10.1007/s13437-023-00313-6

Kneebone, R. (2009). Perspective: Simulation and transformational change: The paradox of expertise. *Academic Medicine, 84*(7), 954. https://doi.org/10.1097/ACM.0b013e3181a843d6

Meyer, J., & Land, R. (2003). Threshold concepts and troublesome knowledge: Linkages to ways of thinking and practising within the disciplines. *Enhancing Teaching-Learning Environments in Undergraduate Courses*, 1–12.

Muckler, V. C. (2017). Exploring suspension of disbelief during simulation-based learning. *Clinical Simulation in Nursing, 13*(1), 3–9. https://doi.org/10.1016/j.ecns.2016.09.004

Sellberg, C. (2017). Representing and enacting movement: The body as an instructional resource in a simulator-based environment. *Education and Information Technologies, 22*(5), 2311–2332. https://doi.org/10.1007/s10639-016-9546-1

Sellberg, C. (2018). From briefing, through scenario, to debriefing: The maritime instructor's work during simulator-based training. *Cognition, Technology & Work, 20*(1), 49–62. https://doi.org/10.1007/s10111-017-0446-y

Sellberg, C., Lindwall, O., & Rystedt, H. (2021). The demonstration of reflection-in-action in maritime training. *Reflective Practice, 22*(3), 319–330. https://doi.org/10.1080/14623943.2021.1879771

Sellberg, C., & Sharma, A. (2024). Toward multimodal learning analytics in simulation-based collaborative learning: A design ethnography of maritime training. *International Journal of Computer-Supported Collaborative Learning*. https://doi.org/10.1007/s11412-024-09435-2

Sellberg, C., & Solberg, M. (2024). Reconceptualizing simulations: Epistemic objects and epistemic practices in professional education. *Philosophy & Technology, 37*(3), 77. https://doi.org/10.1007/s13347-024-00764-6

Wahl, A. M., Giskeødegård, M. F., & Sellberg, C. (2025). Routinization, ritualization and role-play: Comparing nursing, nautical and bioengineering students' perceptions of simulation-based training. *International Journal of Training and Development*. https://doi.org/10.1111/ijtd.12364

Wilson-Thain, S., & Hammond, L. (2023). Examining clinical reasoning through a threshold concept lens. *Health Professions Education, 9*(4) https://doi.org/10.55890/2452-3011.1050

Wyss, C., Rosenberger, K., & Bührer, W. (2021). Student teachers' and teacher educators' professional vision: Findings from an eye tracking study. *Educational Psychology Review, 33*(1), 91–107. https://doi.org/10.1007/s10648-020-09535-z

Susan Harrington is a visiting researcher and former postdoctoral researcher at the Department of Applied Information Technology, University of Gothenburg, specializing in simulator-based maritime education and training. With a PhD in Management Science from the University of Strathclyde, she focuses on scenario-based learning design to support immersive, experiential training in high-risk maritime contexts. Her research bridges pedagogical theory and practical application, examining how scenario exercises can cultivate competencies such as communication, collaboration, reflexive decision-making, and good seamanship through structured debriefings and instructional support. Susan's research also focuses on designing effective learning analytics for simulator-based professional learning.

Facilitating Debriefings for Feedback, Discussion, and Reflection

Amit Sharma

Abstract The process of debriefing is argued to be the most critical stage during the simulation exercises of high-risk maritime operations. Debriefing is an important step in simulation-based learning exercises for promoting structured discussions and identification of learning gaps. However, it is especially crucial when considering the teamwork component of the maritime operations. During the debriefing stage, the participants, i.e., the instructors and the maritime trainees, have the opportunity for group-level reflection and identification of corrective measures for sub-optimal activities in relation to the learning objectives. A variety of mediational tools and resources can be utilized by the instructors for highlighting the performance of the learners and providing feedback. In this chapter, we look into the process of establishing a conducive environment for facilitating the debriefing process, as well as the various debriefing models that can be utilized for orienting the reflections and feedback. The success of the debriefing stage, which occurs towards the end of the simulation exercises, depends to an extent on the productive debates related to the learning objectives and key takeaway messages for the participants. Therefore, the chapter focuses on how the participants can achieve the same, thereby meeting the intended goals of the simulation exercises.

Keywords Simulation-based training · Debriefing · Debriefing models · Maritime Education and Training (MET)

A. Sharma (✉)
Estonian Maritime Academy, Tallinn University of Technology, Tallinn, Estonia
e-mail: amit.sharma@taltech.ee

C. Sellberg (ed.), *Simulations in Maritime Education and Training*, WMU Studies in Maritime Affairs 12, https://doi.org/10.1007/978-3-032-08394-4_5

1 Introduction

Simulation-based learning is one of the common modes of developing the requisite knowledge, skills, and attitude of the trainees in relation to their roles in relatively complex and demanding occupations such as maritime operations. The interactive scenarios practiced in simulation exercises supervised by experienced instructors allow the trainees (hereby referred to as the learners) to learn and practice in a controlled and safe environment. The simulation-based learning experience is usually divided into three distinct stages: (1) Pre-simulation briefing, (2) Simulation exercise (or scenario), and (3) Debriefing (Sellberg, 2018). The previous chapters in this book have already looked into the briefing stage and the intricacies involved when planning and conducting the actual simulation exercises. The present chapter focuses on the debriefing stage, which occurs post-simulation and where the participants discuss, interact, and reflect on their experiences, and which is argued to be a very crucial part of the overall learning activity.

The word "Debriefing" in simple terms refers to a meeting of the participants of a group involved in an activity or operation and reflecting on the performance and overall achievement of the intended aims and objectives. A more formal characterization of debriefing is "a discussion and analysis of experience, evaluating and integrating lessons into one cognition and consciousness" (Gardner, 2013; Lederman, 1992). The antecedents of debriefing are stated to be in the military contexts of the 20th century, specifically during the Second World War, when the post-mission debriefs were used for informational gathering and providing feedback to the involved soldiers. The information gathered was then utilized to improve strategy as a result of the experience and lessons learned (Pearson & Smith, 1985). However, such an approach was characterized as being "top-down" and left little avenue for the involved troops to reflect deeper or share their perspectives on the operations (Gardner, 2013).

Subsequently, the debriefing-related research also emerged from high-risk operational contexts such as aviation, emergency response, healthcare, etc (Dismukes et al., 2006). The focus shifted from gathering information and providing feedback to engaging in dialogue and encouraging participation. The concept of debriefing was further refined by the proponents of these fields to not only associate it with the functional objectives of the operations but also to support group cohesion and provide a supportive climate (Dyregrov, 1997). In the context of education and simulation-based learning, the process is often led by experienced instructor(s) and is particularly geared towards meeting the learning objectives set out at the beginning of the exercises. A successfully conducted debriefing session helps the learners acquire relevant technical and behavioural skills both at the individual and group levels.

Although the debriefing practices vary according to the educational settings and learning objectives related to simulation exercises, Lederman (1992) has described seven key structural elements in debriefing:

(1) Debriefer(s)—often experienced instructor(s) leading the process

(2) Participants—the trainees or the learners for whom the exercise is targeted
(3) An experience—this refers to the simulation exercise or scenario
(4) Impact of the experience—The cognitive or affective component
(5) Recollection—how the experience is recalled by the learners
(6) Report—how the experience is documented (e.g., survey or questionnaire)
(7) Time—The time elapsed post-simulation

The pedagogical premise of debriefing is based on several theoretical foundations related to adult learning experiences. For example, in Kolb's learning cycle, which is characterized by four stage process namely—(1) Concrete experience (2) Reflective observation (3) Abstract conceptualization and (4) Active experimentation, the learner has to ideally engage in each of the stages during the learning process (Kolb, 1984). The simulation scenario can be connected to the first stage of the cycle, i.e., concrete experience, where the learners undergo specific experiences connected to the learning objectives. Post-simulation, when debriefing of the exercise occurs, the learners have the opportunity to reflect on their actions while also linking their experiences to the desired practices. This stage, which is facilitated by the instructor, directly maps to the reflective observation and abstract conceptualization processes in the cycle. The learners, after internalizing the key takeaways, have the opportunity for active experimentation in the future scenarios.

Similarly, the reflective practice theory advocated by Schon (1983) describes the practices of professionals in their field of work, mentioning two types of reflection: (1) Reflection in action and (2) Reflection on action. The debriefing process directly maps to the reflection on action concept, where the focus is on reviewing the outcomes after the action has concluded, and the retrospective analysis of the experience takes place (Sellberg et al., 2021). The conceptual change model given by Lewin and Schein (Lewin, 1947; Schein, 1996) describes the three-stage processes of "unfreezing", "transition", and "freezing" by which conceptual change occurs in learners and new behaviours can be internalized. By confronting the contradictions in the understanding of certain concepts and through guided discussions and reflections during the debriefing, the learners can integrate the lessons correctly into their real-life practice for the future.

It is also worth noting that the term "Debriefing" can be thought of as an all-encompassing word, distinct from "feedback" (Amundsen & Noble, 2025; Karahalil et al., 2023). At times, these words are used interchangeably. However, a distinction amongst them is also recognized in the emergent literature. In simulation-based learning settings, the term "feedback" often refers to one-way information from instructor to student pertaining to improvement of their performance as observed (Sawyer et al., 2016). In contrast, debriefing is argued to be an interactive, bidirectional and reflective conversation or discussion (Voyer & Hatala, 2015). The debriefing conversations can take place between the instructors and the learners. They can also take place within these respective groups, where peers share their perspectives on their experiences. This can allow for the cultivation of both technical and non-technical skills in the context of maritime operations.

2 Debriefing in Maritime Simulation-Based Training and Assessment

Simulation-based training and assessment is an important cornerstone in the domain of maritime education. A vocational approach still remains dominant for preparing seafarers of the future, which relies on specific competences, mentorship from experienced peers/instructors, and teamwork-dependent tasks (Manuel, 2017). Since the debriefing process is often described as the most important stage in the simulation-based learning scenarios and an avenue for group-level discussions and reflections, planning and execution of a productive session is important for meeting the learning objectives (Gardner, 2013: Fanning & Gaba, 2007). Consequently, the role of debriefing becomes central to imparting the prerequisite skillsets for the prospective seafarers and evaluating their readiness for performing high-risk operations at sea.

Rudolph et al. (2008) describe the process of debriefing as a way to conduct formative assessment. It revolves around observing the actions of learners in simulator settings, which depend upon the existing cognitive "frames" held by them. The term frames here refer to schemata or mental models of reality as observed by the learners. These influence the actions taken by the learners when encountering targeted scenarios in simulated settings. The primary tasks of the instructor then become the comparison of the observed actions to desired actions, evaluating the performance gaps, and correcting them as required. Frames are deemed to be invisible but can be ascertained through appropriate queries. Actions engaged by the learners, on the other hand, are observable. The performance gap is the difference between desired actions in the situation and the observed actions. Figure 1 below provides an illustration of the process.

The Standards of Training, Certification and Watchkeeping (STCW) regulations formally lay down the statutory competence requirements for seafarers in merchant shipping. These regulations also enumerate a number of non-technical skills addition to technical skills, that seafarers should demonstrate concerning their job profiles (STCW, 2010). Under these frameworks, maritime professionals participate in simulation-based scenarios to fulfil the requisite knowledge, understanding, and proficiency (KUPs) items of each of the functions listed under their respective roles. More specifically, if we talk about the deck and engine departments, the seafarers undergo dedicated training programs such as *Bridge Resource Management* (BRM) or the *Engine Resource Management* (ERM) courses for the above purposes. To practice the collaborative execution of all the tasks, understanding the role specific

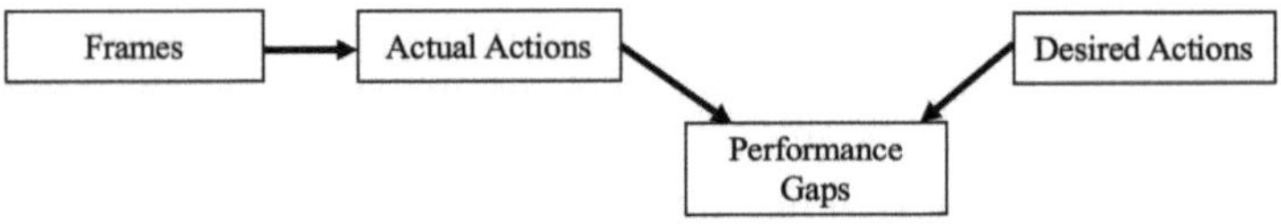

Fig. 1 Debriefing process as formative assessment (Adapted from Rudolph et al., 2008)

requirements and developing correct behavioural approach for the same, a conducive environment during the simulation exercise becomes necessary. The experienced instructors facilitating these simulator sessions attempt to guide the learners by connecting the general lessons learned and challenges observed in the scenarios with a more specific path towards the learning objectives and professional practices onboard (Sellberg, 2018).

When considering simulation exercises involving bridge operations (i.e., involving navigators), a typical approach at the MET institutes is employing full mission simulators that offer a high fidelity, immersive experience akin to real navigation operations at sea (Kim et al., 2021). In such scenarios, the learners are often divided into a small set of teams and stationed at simulators where the scenarios are simultaneously played out. Each simulator has a large field of view and replicas of all the bridge instruments. The instructor usually monitors the progress of the exercise from the monitoring and control station. After the simulation exercise concludes, the participants often gather back at a dedicated area to take part in the debriefing session. Figure 2 below provides an example illustration of how a typical debriefing session is conducted in maritime domain.

Typically, video playbacks are utilized on a large screen area in front of the whole group to go through the exercise and discuss it in greater detail. This data is also accompanied by playbacks from bridge devices such as RADAR or ECDIS when possible. Alternatively, the instructor may use snapshots of the images from a section of the exercise and integrate with their own comments before presenting it in the form of PPT for wider discussions. Use of video playbacks or PPTs can help in highlighting actions taken by each team and the instructors try to connect these actions with the overall learning objectives. More recently, there is also the possibility of using physiological outputs such as eye-tracking data to assist in debriefing processes (Haavardtun et al., 2025). Technological aids such as these help in orienting the discussions to a common frame of reference for the participants and providing objective information to a degree. However, the success in meeting the learning

Fig. 2 An example of debriefing session at MET institute. Image credit – Charlott Sellberg

objectives depend to a significant extent on the ability of the instructor in organizing the learning activities meaningfully for the learners (Sellberg, 2018).

3 Different Debriefing Models for Promoting Structured Discussion, Reflections, and Feedback

While the debriefing process is described to an extent in the literature associated with maritime simulation-based training and assessment, relatively less information is provided with respect to the available models and frameworks that can be followed for ensuring structured discussion, reflections, and feedback. A number of models, frameworks, and guidelines are, however, available in the wider area of simulator-based learning state-of-the-art. While most of these models have their origins in clinical or healthcare practices, they can still be used for maritime debriefing, due to the core principles of simulation being common across similar high-risk domains (Sellberg & Giskeødegård, 2025). The use of a particular model or framework naturally depends on factors such as the type of simulator utilized, the expertise of the instructor, time duration, and so on. Furthermore, various models exhibit certain similarities (as well as their unique characteristics/dissimilarities) when describing the debriefing process. The phases of these models or frameworks typically range from three to as many as six or seven in some cases. Typically, they are grouped as either a three-phase model or a multiphase model (more than three). Several research articles (Fanning & Gaba, 2007; Sawyer et al., 2016) have described various models and frameworks employed for debriefing purposes. In the section below, some of the prominent debriefing models are described, which can be employed to conduct a structured debriefing session.

3.1 Reaction - Analysis - Summary

The three-phase debriefing method of reaction, analysis, and summary is given by Rudolph et al. (2006, 2008) and has its origin in clinical practices. In this method, the debriefing conversation begins with the "Reaction" phase, where the instructor has to first allow the learners to express themselves for a brief uninterrupted period, where the initial response to the simulation exercise will be registered. This phase can be initiated by the instructor by first asking an open-ended question, such as—"*How did it go?*" or "*How are you feeling?*". It is common for the learners to express positive or negative emotions associated with the exercise here. However, the instructor doesn't need to be passive in this phase, and he/she can guide the class by validating the feelings of the learners and presenting facts to alleviate any confusion. The subsequent "Analysis" phase is where the majority of the debriefing dynamics take place (Lymer & Sjöblom, 2024; Skovholt et al., 2019).

In addition to using any technological tools available for this phase to aid the instructor and participants in revisiting the simulation exercise, the focus in the Analysis phase is directed towards generalizing the lessons learned and connecting them to the actual practices at sea. Rudolph et al. (2008) give four step process of facilitating formative assessment specially tailored to the analysis phase of the method – (1) Observe gap between desired and actual performance (2) Provide feedback about the gap (3) Investigate the basis of the gap and (4) Assist in closing the gap through didactics and discussions. An advocacy-inquiry approach is the most common method utilized by the instructors to structure and present their feedback to the learners and encourage them to respond. This helps instructors to deliver non-judgmental, constructive criticism (Rudolph et al., 2006). Primarily, this begins by first stating clearly what was observed by the instructor using phrases such as—*"I saw… ", "I observed… "* etc. This minimizes guesswork in the classroom and brings both instructor and the learners to common frame of reference with respect to what objectively occurred in the session. It is followed by advocacy statements where the instructor has the opportunity to deliver the feedback and correct the observed deficiencies in a non-threatening, neutral tone. The instructor can utilize phrases such as *"I think... ", "I suggest… "*, and similar to orient the learners to the correct behaviours and meet the overall objectives of the exercise. However, to uncover the assumptions of the learners and further promote dialogue and reflections, the instructor should also use inquiry statements to allow the learners to share their perspectives. The instructor can use phrases such as "*I am curious…*", "*I wonder… "* and then allow the learners to state what was going on in their minds which led to the particular decisions.

In the "Summary" phase, as the term suggests, the participants and the instructors review the lessons to be learned from the whole episode. The purpose of this phase is to synthesize the key takeaways and formulate "rules of thumb" if possible. The instructor can begin this phase by using phrases such as "What went right?", "What will you do differently?" and similar to allow the learners to appreciate the information they have understood and uncovered in the last two phases in the debriefing. The instructor can assist the learners by validating their thoughts and connecting them to the overall learning objectives. The aim of the instructor should be to remind the learners of the primary purpose of the simulation exercise and provide suggestions for future course of action.

3.2 Defusing - Discovering - Deepening

The Defusing-Discovering-Deepening model given by Zigmont et al. (2011) is another prominent model used for conducting debriefings. This model is sometimes abbreviated as the "3D model". The model is based on Kolb´s experiential learning cycle, adult learning theory, and learning outcomes model, and is tailored for assisting the instructors in conducting debriefing sessions post-simulation.

The first phase, as the term suggests, is about allowing the learners to express themselves immediately after the simulator session is concluded and allowing for first impressions to be described. The instructors should use subtle prompts to initiate the discussions. It is not necessary or required for every learner to speak, especially if the group is large. However, every learner should have the opportunity to speak in this phase. The instructor has to ensure that an orderly and safe environment is present at the start of the debriefing. It is also possible for 1–2 students to take the lead from their respective groups and narrate their experiences after a small internal discussion. The important point is that all of the learners converge towards a shared understanding of what occurred during the simulation. The objective of this phase for the instructor, in addition to helping the learner express their emotions, is to recognize the needs of the learners and what key points are particularly important for them. It is possible that the learners might overshoot their thoughts during this phase and may start premature analysis of the simulation scenario in this phase only. The instructor has the responsibility to gently guide the discussions back towards allowing the expression of feelings first. Another objective for the instructor in this phase is to acknowledge and validate the initial impressions of the learners.

Subsequently, the "Discovering" phase is where the reflections can begin for the learners. The instructor can aid this process by recalling the steps taken by the learners during the simulation in an objective manner and inviting them to comment on their reasoning, thereby initiating the reflective dialogues. If available, video playback of the exercise can also be utilized. These steps directly map to the "reflective observation" part of Kolb´s learning cycle. Thereafter, the instructor should guide the dialogue to identify the mental model of the learner. In other words, ask them in a polite non non-judgmental way the reason for their taking a particular action. This is the central objective of the discovery phase, and it is a vital step before any learning can occur. The instructor can offer a "correct" mental model if necessary, describing the expected actions that should be taken in the particular scenario, thereby highlighting the learning gap existing which should be covered subsequently. During these steps, the "abstract conceptualization" part of Kolb´s learning cycle has been initiated in the learner's mind as he/she tries to compare their performances and make structural connections of different associated concepts.

In the "Deepening" phase, the learner strives to make meaning of the information gathered in the prior phases and test it in novel situations. This phase directly maps to the active experimentation stage of Kolb's learning cycle. Ideally, this can be achieved through a second round of simulation, where the learner now has the opportunity to test out the new approach. However, it is possible that due to logistics or time constraints, a second round of simulation is not immediately possible to execute. In such cases, instructors should pose queries to the learners regarding how they plan to apply the lessons learned from the prior phases to novel scenarios and perform a walkthrough. The overall goal of the "deepening" phase of the debriefing is to make explicit the connection between the target mental model internalized by the learner to the external environment and actual practices performed onboard.

3.3 *Description - Analysis - Application*

The Description-Analysis-Application model given by Jaye et al. (2015) is another three-phase model utilized for performing debriefing sessions post-simulation. It is also known as the "Diamond Debrief" model owing to the diamond-shaped flowchart aid described in the framework, which is used to conduct the debriefing in a structured manner. This model is argued to be more tailored for facilitating the exploration of non-technical skills during the simulator exercise. It is also markedly different from the previous models mentioned, as it provides emphasis on emotional aspects in a different order during the debriefing process. The authors acknowledge this as a conscious decision due to the cultural settings where the model was developed.

The "Description" phase, as the word suggests, is primarily focused on describing in detail what happened during the simulation scenario. This phase is actively led by the instructor and consists of a step-by-step walkthrough of the whole exercise. The phrases such as "*what happened next...*" and similar are used to nudge the dialogue forward while the sequential unfolding of the scenario. Despite explicit avoidance of emotional aspects of the simulation, the model does emphasize creation of a safe debriefing environment and development of a shared context in the description phase. Certain transition phrases such as "*This scenario had the aim of...*", "*What should we do here?*" or "*Is everyone following?*" can be used to gently guide the learners to the next phase of the model i.e., "Analysis".

The "Analysis" phase is where most of the time is usually spent in this framework of debriefing. In the analysis phase, the instructor can now ask certain questions related to the emotional state of the learners, such as "*How did you feel?*" and similar. These are to be accompanied by queries such as "*Why?*" and appropriate periods of silence, where applicable, to help the learners articulate themselves. The instructor should guide the discussions towards the non-technical aspects of the scenario and go through them step by step with the learners. Focus should be given towards making sure that the l earners grasp the operational requirements beyond technical skills and can understand the need for timely and correct communication, decision making, and collaborative task execution. The instructor should strive to keep the dialogues positive and productive. This phase can be concluded with phrases such as "*So, what have we learned in this scenario?*" or "*What do we agree to do in this situation?*" before moving on for the "application" phase. For the transition, the instructor can try to repeat the learner´s answers and allow them to correct or amend if required.

The "Application" phase of the diamond debrief model is concerned with clarifying with the learners how they may apply the lessons learned from the simulation scenario to their own practices. In this phase, generally, any alternative interpretation is not encouraged, and the discussions are focused on concrete, real-life examples. The instructor can ask one or two members of the group to summarize the key points and contextualize the non-technical skills targeted in the scenario. The primary takeaway for the learners should be a personalized yet specific set of actions that should be undertaken for the situation encountered during training.

3.4 TeamGAINS

The abbreviation TeamGAINS stands for—Guided team self-correction, Advocacy-Inquiry, Systemic-Constructivist. It's a multi-phase debriefing model given by Kolbe et al. (2013) with a focus on team training. In simple terms, it is conceptualized as a combination of best practices found by the authors from the available debriefing literature and builds on guided team self-correction, advocacy-inquiry, and systemic-constructivist methods of debriefing. This entails providing a non-judgmental environment to the learners where self-correction can take place at the team level, with specific feedback and reflections offered by the instructor. Furthermore, the focus is given towards interactions and relationships between the team members rather than individual actions. It consists of six steps that should be followed in sequence during the debriefing process:

1. **Reaction**—Congruent to some of the models mentioned earlier, this step requires that the instructor announces the start of the debrief and allows without interruption for each of the participants to voice their emotions. This can be done by using a simple phrase such as "*So, how do you feel?*". The instructor can also note which areas were reported by the learners to be particularly important for them. Towards the end of this round, the learners should feel ready for reflection.
2. **Debriefing of the scenario**—This step involves a step-by-step walking through of the specific targeted scenario. The instructor takes the lead in performing this step and is using a narrative approach, ask questions such as "*What happened here*" and allowing the learners to answer. It is important to go through each of the involved steps of the scenario and make gentle corrections to the observed actions when necessary. The instructor can use a personal story or adapt a first-person voice and describe to the learners what actions he or she would have taken. Towards the end of this step, the clarity with regards to the specific scenario should be established.
3. **Transfer from simulation to reality**—The instructor should subsequently try to connect the scenario enacted in the simulation to the real world. The instructor can ask the learners about any past experiences they might have related to this scenario. The goal is to encourage the learners to think about real-life applications of the lesson.
4. **Discuss the behavioural skills**—In this step, the expert should dive deeper into specific behavioural skills that are expected from the learners. The instructor can try to connect the observed actions of the learners to the overall learning objectives of the exercise. Another approach to facilitate this step is to use the technique of "circular questioning," which involves a third individual to describe the relationship and roles between two other members of the team. This removes any

ambiguity regarding the roles and expectations of team members. Furthermore, video-playback of the exercise or multi-media tools (when available) are used in this step to dive deeper into the actions expected from the learners.

5. **Summary**—As the name suggests, the instructor summarizes the debriefing session in this step. This can be done by the instructor by using phrases such as "*So, what did we learn from the scenario?*" or "*If another learner had observed you, what might they have learn?*".
6. **Supervised practice**—The final step can be optional but is recommended. The summary can immediately be followed up by the practice of steps that were suboptimal by the learners, with guidance from the instructor. The instructor can also lead this step by themselves, performing the exercise and demonstrating correct actions to be executed. Alternatively, additional dialogue regarding what could have been done differently for the scenario is carried out, in case the time and logistics are not available for a short repeat of the simulation.

3.5 DEBRIEF

The DEBRIEF model is an adapted version of *After Action Review* (AAR) framework from the US military, tailored for simulation debriefing by Sawyer and Deering (2013). It provides a clear focus on the learning objectives and pre-identified performance standards. Certain steps along the actual debriefing process should be carried out when using this framework, basis the original AAR guidelines—planning, preparation, conducting the actual debriefing, and follow-up. The debriefing process, as described in the DEBRIEF model, consists of seven steps, with each of the letters of the model contributing to the abbreviation. Steps 1–4 should be actively led by the instructor, whereas steps 5–7 could be facilitated indirectly if needed.

1. **Define rules**—Setting the stage before beginning the debriefing process can make it efficient overall. The instructor should make sure to notify the learners that debriefing is not a critique of their actions. Furthermore, it should be made clear that it is intended as a reflective discussion rather than one way feedback. If there are time constraints on how much time can be spent on this stage, it should be clarified. The instructor should make sure to set some ground rules for a conducive discussion and a non-threatening, professional environment.
2. **Explain learning objectives**—It is important for the learners participating in the simulation exercise to know the learning objectives. The instructor should explicitly state them when starting the debriefing session to promote engagement with the discussions that follow and the development of shared mental models in all the learners. They should all be on the "same page". The learning objectives can also be stated even before the start of the simulation, i.e., during the briefing stage, and repeated once more before the start of debriefing.
3. **Benchmark for performance**—It is of vital importance to mention clear performance benchmarks and standards against which the learners are being

evaluated to avoid ambiguity and promote psychological safety amongst the participants. It is possible for the instructor to integrate these performance benchmarks with the learning objectives beforehand. The key tasks involved during the exercise and the associated standards for successful completion and learning should be defined.

4. **Review expected actions**—The instructor should begin the debriefing by first clarifying what an ideal outcome could have been. It is not necessary to go into minute detail here, but the learners should get an abstract idea of what was expected from the simulation scenario. The instructor can use phrases such as "What we intended to happen in this scenario…" or similar.
5. **Identify what happened**—As the name suggests, the next steps are about determining what transpired during the simulation exercise. This entails the instructor asking a series of open-ended questions and allowing the learners to narrate the occurrences. It is mentioned for this step that the instructor has to act as a "cognitive detective" to uncover the frames used by the learners and use a series of advocacy-inquiry queries to ascertain the causes for the actions undertaken.
6. **Examine why**—Practically, this step takes place simultaneously with step 5. With follow-up queries where necessary, the instructor has to allow the learners to describe the reasoning behind their actions. If necessary, the instructors should document the reasoning given by the learners, so that any error in setting up the whole scenario or any omitted information at the beginning can be rectified.
7. **Formalize learning**—Finally, in the final step of the framework, the instructor should summarize the complete exercise at the group level for the learners. The instructor should mention—what went well, what did not go well, and what actions can be taken in the future. The debriefing should conclude with clear take-home messages for the learners.

As can be seen above, several models and frameworks exist in the literature for conducting debriefing post-simulation. However, it should also be noted that the models and frameworks primarily act as guidelines, and instructors should focus more on making the process of reflection and discussion easier for the learners (Sellberg, 2018).

4 Evaluating the Debriefing

Since debriefing is a process that will be conducted repeatedly in the educational program for maritime simulator-based training, it is important that the process itself is also documented and evaluated, if possible, for generating actionable insights and possible improvements. An approach for this purpose could be the use of recognized and validated survey scales tailored for debriefing evaluation. The scores generated by the use of such instruments can help quantify the performance of the instructor or the experiences of the learners, which can readily be compared with other debriefing sessions when required. There are several standardized scales available in the research literature that are regularly used for the same purposes.

The Debriefing Experience Scale (DES), developed by Reed (2012), measures the perception of the learners in terms of educational utility and psychological safety during the debriefing process. It is a validated 20-item scale where the respondents rate each item on a scale of 1 to 5, from strongly disagree to strongly agree. DES has four sub-scales or domains that are measured: (1) Learning and making connections, (2) Facilitator skill in guiding reflection, (3) Meaningful learning, and (4) Recollection of the experience. The DES is administered immediately after the debriefing session, and the obtained results can be used for evaluating the quality of debriefing longitudinally.

The Objective Structured Assessment of Debriefing (OSAD) developed by Arora et al. (2012) is another validated scale that is utilized for assessment of debriefing. OSAD is used by experienced researchers or faculty peers to evaluate the instructors conducting the debriefing. It measures their performance in eight areas, namely—(1) Approach (2) Learning environment, (3) Engagement, (4) Reaction phase (5) Description phase (6) Analysis phase (7) Summary phase, and (8) Performance improvement focus. Similar to DES, a five-point Likert scale rating is used from poor (1) to excellent (5). OSAD is typically used for quality control or faculty development purposes.

While DES or OSAD have their origins in healthcare simulation, they can be adapted for use in maritime simulation debriefing, if the clinical terminology is substituted with maritime terminology, where applicable. Similarly, another scale known as Debriefing Assessment for Simulation in Healthcare (DASH) by Simon et al. (2010) is commonly used for assessing the behavioural skills of the instructor and the quality of debriefing. It has six main elements: (1) Establishes an engaging learning environment, (2) Maintains an engaging learning environment, (3) Structures debriefing in organized version (4) Provokes engaging discussions (5) Identifies and explores the performance gaps and (6) Helps trainees achieve or sustain good future performance. The seven-point Likert scale is used for rating each of the elements from (1) Extremely ineffective to (7) Extremely effective. The DASH instrument has three versions each: a student version for the learners to give feedback on the debriefing experience, a rater version for the researcher or peer to assess instructor debriefing skills, or a third instructor version for self-assessment. DASH requires some prior experience and training for its administration in educational settings; however, its predominant neutral wording makes it suitable for use across other high-risk simulation scenario debriefing, especially for the maritime domain.

The different models and frameworks mentioned above focus on different aspects of debriefing. The use of any particular model or framework is naturally dependent on several factors, such as the purpose of the simulation exercise, the type of simulator used, time duration, class size, to name a few. It is important for the maritime instructors engaged in the debriefing process at their respective institutes to make their own tailored toolbox for conducting a productive debriefing session. This can be done by taking into account the utility of any particular approach with the resources available at their disposal. The instructor or other active stakeholders in the institute should take it into account if they plan to use a similar approach multiple times to ensure a consistent benchmark for longitudinal comparisons. The technical resources available at their disposal will also influence which method(s) can prove particularly suitable in meeting the overall needs of the educational program. Additionally, continuous professional development of the instructors themselves, over a period of time, is important to ensure the success of debriefing sessions in the educational program utilizing simulators.

5 Good practices for Debriefing

A structured and productive debriefing session post-simulation is critical for achieving the full benefit of the exercise and turning the experience into long-term lessons for the participants. As illustrated in the above sections, some good practices should be observed for organizing the debriefing session to meet these objectives:

- Debriefing process should be planned right from the beginning and together with the overall logistical arrangement of the simulation exercise. Adequate resources and time should be allocated for conducting it.
- Debriefing should be conducted in a conducive environment. It is paramount that the psychological safety of the participants is ensured.
- Debriefing process should be adequately structured and follow a theoretical framework where possible.
- Debriefing should be conducted by experienced instructor(s), ideally trained for conducting the session. The instructor(s) should be able to devote proper attention towards the debriefing process and provide constructive feedback where necessary. It is important to engage the learners through active dialogue throughout the process.
- Debriefing process should be documented and evaluated for potential improvement. A dedicated researcher or observer can assist the instructors in doing so. The feedback can be informal or verbal in some cases. However, it is better to use validated survey instruments where possible for generating actionable insights.
- Debriefing should conclude with clear takeaway messages for the learners. It should align with the pre-set learning objectives of the exercise.

6 Conclusion

The aim of this chapter was to provide the stakeholders involved with maritime simulation-based learning activities with a roadmap to bridge theory with practice. It is important to tie the debriefing session to the overall learning objectives of the simulation exercise and provide psychological safety for the learners to engage without any reservations. The current cognitive frames of the learners should be elicited and ideally transformed into updated mental models. There are several mediational tools available for supporting the instructors in conducting debriefing, but the most important resources are their own skillset and communication strategies. A structured theoretical framework is also required to maximize the effectiveness of the debriefing using established mechanisms. Use of standardized and validated scales, for example, can help with generating actionable insights and potential improvement of the existing practices.

References

Amundsen, T., & Noble, A. (2025). Are you simulator ready? A study of maritime communication competence in the full-mission bridge simulator. *WMU Journal of Maritime Affairs*, 1–17.

Arora, S., Ahmed, M., Paige, J., Nestel, D., Runnacles, J., Hull, L., et al. (2012). Objective structured assessment of debriefing: bringing science to the art of debriefing in surgery. *Annals of Surgery, 256*(6), 982–988.

Dismukes, R. K., Gaba, D. M., & Howard, S. K. (2006). So many roads: Facilitated debriefing in healthcare. *Simulation in Healthcare, 1*(1), 23–25.

Dyregrov, A. (1997). The process in psychological debriefings. *Journal of Traumatic Stress, 10*, 589–605.

Fanning, R. M., & Gaba, D. M. (2007). The role of debriefing in simulation-based learning. *Simulation in Healthcare, 2*(2), 115–125.

Gardner, R. (2013). Introduction to debriefing. *Seminars in Perinatology, 37*(3), 166–174.

Haavardtun, P., Komandur, S., Lunde, P., Mjelde, F. V., Pignoni, G., Sharma, A., & Volde, F. (2025). Cognitive workload, eye-tracking and communication: A mixed method study in a navigational training context. *Necesse, 8*(1), 94–117.

Jaye, P., Thomas, L., & Reedy, G. (2015). 'The Diamond': A structure for simulation debrief. *The Clinical Teacher, 12*(3), 171–175.

Karahalil, M., Lützhöft, M., & Scanlan, J. (2023). Formative assessment in maritime simulator-based higher education. *WMU Journal of Maritime Affairs, 22*(2), 181–207.

Kim, T. E., Sharma, A., Bustgaard, M., Gyldensten, W. C., Nymoen, O. K., Tusher, H. M., & Nazir, S. (2021). The continuum of simulator-based maritime training and education. *WMU Journal of Maritime Affairs, 20*(2), 135–150.

Kolb, D. A. (1984). *Experiential learning: Experience as the source of learning and development.* Prentice-Hall.

Kolbe, M., Weiss, M., Grote, G., Knauth, A., Dambach, M., Spahn, D. R., & Grande, B. (2013). TeamGAINS: a tool for structured debriefings for simulation-based team trainings. *BMJ Quality & Safety, 22*(7), 541–553.

Lederman, L. C. (1992). Debriefing: Toward a systematic assessment of theory and practice. *Simulation & Gaming, 23*(2), 145–160.

Lewin, K. (1947). Group decision and social change. *Readings in Social Psychology, 3*(1), 197–211.

Lymer, G., & Sjöblom, B. (2024). Interaction in post-simulation debriefing. *Learning, Culture and Social Interaction, 48*, 100855.

Manuel, M. E. (2017). Vocational and academic approaches to maritime education and training (MET): Trends, challenges and opportunities. *WMU Journal of Maritime Affairs, 16*, 473–483.

Pearson, M., & Smith, D. (1985). Debriefing in experience-based learning. In *Reflection: turning experience into learning.* (pp. 69–84). Routledge.

Reed, S. J. (2012). Debriefing experience scale: Development of a tool to evaluate the student learning experience in debriefing. *Clinical Simulation in Nursing, 8*(6), 211–217.

Rudolph, J. W., Simon, R., Dufresne, R. L., & Raemer, D. B. (2006). There's no such thing as "non-judgmental" debriefing: A theory and method for debriefing with good judgment. *Simulation in Healthcare, 1*(1), 49–55.

Rudolph, J. W., Simon, R., Raemer, D. B., & Eppich, W. J. (2008). Debriefing as formative assessment: Closing performance gaps in medical education. *Academic Emergency Medicine, 15*(11), 1010–1016.

Sawyer, T., Eppich, W., Brett-Fleegler, M., Grant, V., & Cheng, A. (2016). More than one way to debrief: A critical review of healthcare simulation debriefing methods. *Simulation in Healthcare, 11*(3), 209–217. https://doi.org/10.1097/SIH.0000000000000148. PMID: 27254527

Schein, E. H. (1996). Kurt Lewin's change theory in the field and in the classroom: Notes toward a model of managed learning. *Systems Practice, 9*, 27–47.

Schon, D. A. (1983). *The reflective practitioner: How professionals think in action*. Basic Books.

Sellberg, C. (2018). From briefing, through scenario, to debriefing: The maritime instructor's work during simulator-based training. *Cognition, Technology & Work, 20*(1), 49–62.

Sellberg, C., & Giskeødegård, M. F. (2025). Learning from simulation: Pedagogical principles and presuppositions in nursing and nautical education. *Journal of Vocational Education & Training*, 1–15.

Sellberg, C., Lindwall, O., & Rystedt, H. (2021). The demonstration of reflection-in-action in maritime training. *Reflective Practice, 22*(3), 319–330.

Simon, R., Raemer, D. B., & Rudolph, J. W. (2010). Debriefing assessment for simulation in healthcare. *PsycTESTS Dataset.*

Skovholt, K., Nordenström, E., & Stokoe, E. (2019). Evaluative conduct in teacher–student supervision: When students assess their own performance. *Linguistics and Education, 50*, 46–55.

STCW. (2010). *International convention on standards of training certification and watchkeeping for seafarer*. IMO.

Voyer, S., & Hatala, R. (2015). Debriefing and feedback: Two sides of the same coin? *Simulation in Healthcare, 10*(2), 67–68.

Zigmont, J. J., Kappus, L. J., & Sudikoff, S. N. (2011). The 3D model of debriefing: defusing, discovering, and deepening. *Seminars in Perinatology, 35*(2), 52–58.

Amit Sharma currently works as an Assistant Professor at the Estonian Maritime Academy, Tallinn University of Technology, where he leads the Nautical Sciences Research Group. He has a professional background as a navigation officer with 7 years of working experience at sea. His educational qualifications include a PhD in nautical operations and an MSc in maritime management with technical specialisation from the University of South-Eastern Norway. His research interests are maritime human factors, autonomous shipping, simulation-based training, and assessment. He has published several articles and reports on these topics from a practitioner's viewpoint.

Assessing Competence Through Simulation

Charlott Sellberg

Abstract This chapter explores the challenges associated with using maritime simulators for assessing seafarer competence, with a particular focus on the validity, reliability, and fairness of such evaluations. Key issues addressed include the potential for assessor bias and the inherent complexities of evaluating professional knowledge as it is enacted in dynamic, simulated environments. The chapter emphasizes the need for standardized assessment criteria, focused assessor training, and strategies to mitigate subjective evaluation bias. Drawing on insights from multiple studies, the chapter offers practical guidelines for enhancing the effectiveness of simulation-based competency testing. By bridging the gap between theoretical assessment principles and the practical realities of maritime training, the chapter provides a roadmap for developing fair and reliable assessment practices that reflect the challenges of evaluating professional knowledge in action.

Keywords Simulation-based assessment · Performance assessment · Competence assessment · Assessment models · Maritime Education and Training (MET)

1 Introduction

In the last two decades, maritime education has moved, at least to some extent, from written exams to competency assessments based on practical scenarios in simulators. At the international level, the International Convention on Standards of Training, Certification and Watchkeeping for Seafarers (STCW) regulates the training and certification of seafarers to increase the *validity* and *reliability* of training and assessment in the global maritime sector (International Maritime Organisation,

C. Sellberg (✉)
Department of Applied Information Technology, University of Gothenburg, Gothenburg, Sweden
e-mail: charlott.sellberg@ait.gu.se

C. Sellberg (ed.), *Simulations in Maritime Education and Training*, WMU Studies in Maritime Affairs 12, https://doi.org/10.1007/978-3-032-08394-4_6

1978). According to STCW, each authority must ensure that instructors, supervisors, and assessors have the required competence for the specific type and level of training or assessment of seafarers that the convention prescribes, emphasizing the importance of assessing whether the execution of various operations is carried out professionally and correctly, as well as the importance of assessing so-called non-technical skills. Moreover, for assessing skills in a simulator environment, STCW prescribes that assessors must possess adequate knowledge and understanding of the competencies being evaluated, i.e., they must be qualified for the specific tasks they are assessing. In addition, the assessor should receive guidance in assessment methods and practices and have acquired practical assessment experience under the supervision of an experienced assessor (STCW Convention 1978, Manila Addendum). Thus, the STCW Convention emphasizes both the professional and pedagogical competence of simulator instructors.

Furthermore, national and local university policies generally emphasize that examinations should be designed and carried out in a legally secure and fair manner, highlighting core principles such as validity, reliability, and fairness. This creates several challenges for instructors working with assessment and certification in a simulator environment. Simulation-based tests must meet high standards for assessment and certification formulated in international, national, and local policies. On the other hand, the tests should be adapted to the complexity that characterizes the training situation. How knowledge in action can be assessed is a classic question in both pedagogical research and pedagogical practice and one of the greatest challenges in maritime education. The aim of this chapter is to investigate the challenges surrounding assessment in a simulator environment. Specifically, we will focus on how the practical implementation of simulator-based competency tests meets higher education requirements and the profession's regulations regarding valid, reliable, and fair assessments of professional competence.

2 Striving for Valid, Reliable, and Fair Assessments

2.1 Validity of Simulation-Based Assessments

There are some basic concepts and core principles regarding assessment in higher education in both policy, research, and educational practice. One of these concepts is *validity*, a concept that highlights that assessments must measure what they are intended to measure, i.e., the relevance of the measurement (Brennan, 2023). Measures should align with the learning outcomes and objectives of the course or program. Valid assessments accurately reflect the knowledge, skills, and abilities that are being tested. As stated in standards published by American Psychological Association (APA) in collaboration with American Educational Research Association (AERA) "validity refers to the appropriateness, meaningfulness and usefulness of the specific inferences made by test scores" (Brennan, 2023, p. 2). In general terms, it has been said that assessments of performance should (a) align to the curricular or professional learning objectives, (b) be authentic, direct assessments of a student's performance, and (c) align to the discipline's standards of what constitutes good practice or competence (Conn et al., 2020 p. 100843). A common approach to ensure validity of simulation-based assessment is to involve subject matter experts to evaluate scenarios and assessment models for realism and accuracy (Mirza et al., 2020). This helps ensure the scenario reflects realistic professional situations and the assessment criteria aligned to professional best practices.

To ensure that simulation-based tests of competency are conducted in a manner that ensures validity in assessments, several studies have been conducted in Maritime Education and Training (MET) over the years. In a study from 2008, Emad and Roth argue that the implementation of simulator-based assessment in compliance with the STCW Convention has changed the learning objectives in MET. Emad and Roth (2008) argue that instead of focusing on work-relevant tasks, training became focused on passing competence tests in compliance with the Convention. As a result, this change in learning objectives has led to the use of assessment methods that fail to measure the professional skills that enable students to put to use their competence gained from MET in workplace contexts onboard ships. In other words, Emad and Roth (2008) raises critical concerns whether the STCW Convention has increased or decreased the validity of simulation-based tests of competence. On the other hand, in a study from 2011, Gekara and his colleagues argue that simulator-based assessment has a distinct focus on maritime operations and professional aspects of navigation. Such aspects include, for example, maintaining vessel course and speed, safe distance from other vessels, and the required draft. Gekara et al. (2011) argue that this focus is problematic because it prevents from providing training and assessing the so-called higher cognitive skills, such as comprehension, application, analysis, synthesis, and evaluation, which is also emphasized in STCW. Moreover, Gekara et al. (2011) argue that the scenarios in the simulation-based tests of competence is very similar to the scenarios carried out during training. Based on these observations, Gekara et al. (2011) conclude that the strong focus on maritime operations in the current MET system favors "examination coaching" and "rote learning" rather than high-quality training or the effective evaluation of "essential knowledge and skills" (p. 98). Even if a maritime operation might seem rather straightforward for an outsider, such as crossing a TSS lane or going to port, conducting maritime operations in a proficient, professional, and safe manner requires advanced knowledge of the vessel, the marine environment, and the traffic situation at hand as well as a high level of cognitive effort (Sellberg & Lundin, 2018). For this reason, the professional competence of experienced seafarers is essential to ensure a valid assessment of skills for maritime certification and course examination since the assessments relies heavily on their professional intersubjectivity—the ability to interpret and understand the reasoning and actions of others in a professional context (Sellberg & Lundin, 2018). In this process, the simulator instructor's subjective professional judgement is an absolute prerequisite for conducting assessments of professional performance. As formulated by one of the instructors interviewed in Sellberg (2020):

> I think that... On one hand measurability... There is justice in measurability. And on the other hand, we have the instructors' decisions. Did this go well or didn't it go well? I understand the eagerness to be fair and the pathos for justness, it's great in many respects. But at the same time, it's like... You have a certain kind of knowledge, a sense where you can assess the student. We need to make this as fair as possible, but we need to keep these aspects of being skillful, being a human and having this sense of... It's not really in line with how science is constructed in a technical university, but it's still important.

2.2 *Reliability of Simulation-Based Assessments*

Besides validity, reliability is another core principle for conducting assessment in higher education. Reliability highlights how assessments should produce *consistent* results over time and across different contexts (Brennan, 2023). Reliable assessments ensure that student performance is measured accurately and fairly, without being influenced by extraneous factors. A key aspect of reliability is *inter-rater reliability*, which refers to the consistency of scores given by multiple assessors when evaluating the same performance (Reid et al., 2012). It ensures that different raters produce similar scores, contributing to the overall fairness and accuracy of the assessment process.

Throughout a series of studies based on video data of practical assessment situations in maritime education, Sellberg emphasize the importance of continuing scrutinizing questions of *what* to assess and questions about *how* knowledge-in-action should be assessed to increase reliability during simulation-based tests of competence. Sellberg et al. (2019) is the first of this series of studies analyzing the implementation of competency tests in a simulator environment. The results of the study show that the students who are at the beginning of their education when the proficiency test is carried out are still not able to fully master the complex task, they are faced with during the competence test. The analysis also shows that the instructor who assesses the task is sensitive to the students' difficulties and supports them in various ways in completing the competence test. On the one hand, such strategies can be seen as a threat to the reliability and consistency of assessments. On the other hand, the instructors' support is sometimes a prerequisite for the student to be able to be able to complete the scenario and thus, to be able to demonstrate their competence. Educational research from science education has showed that teacher support during assessment tasks help students to express their ideas, thoughts, and reasoning, and thus displaying their strengths and weaknesses in a way that might increase the validity of the assessment (Kang et al., 2014). Hence, instructional support during competence tests is not a simple matter of right or wrong. Rather, support is to be implemented with good judgement and reflections about fairness during competence tests.

The results that the instructors provide different kinds of support to students during competence tests in the simulator are confirmed in Sellberg et al. (2022) where analyzes of different ways of asking the same question to students during competency tests are in focus. The results show how the instructor supports the students to show what they can do by using strategies such as reformulating questions or clarifying the students' partially correct answers. In this work, the navigation equipment in the simulator environment was an important resource for anchoring abstract questions in concrete working methods on the bridge. Although these strategies can be seen as a threat to the reliability and fairness of assessments, Sellberg et al. (2022) argue that it's unrealistic to think that assessors and students act completely independently. Rather, the assessment is focused on evaluating a student's ability to engage in the practices of becoming a maritime professional. As a result, Sellberg et al. (2022) suggest that the process is more intertwined: assessors play a role in guiding students, giving feedback, and shaping their understanding of what it means to be part of the maritime community. This means that the assessors' involvement affects how students learn and demonstrate their skills, rather than simply standing back and judging them from a distance. Sellberg et al. (2023) explores competence tests in the simulator environment during the students' last year of education and shows how these educational events both play an important function in assessing the students' professional skills and as a tool to strengthen the students in their professional identity when they are about to leave the educational programme for their first employment at sea. By endorsing the students' professional competency, as well as sending them off into the future with a handshake, the assessor shows how the students through their displayed competence now are part of the community of professional mariners and are to be treated as such. To sum up, the findings from these studies shows how competence tests in a simulator environment have strong formative elements, being part of the overarching goal of educating future seafarers that will uphold the high standards of safe and professional conduct that we refer to as good seamanship. While this close connection between formative and summative assessment might complicate the reliability of assessments made during simulation-based tests of competence, it is important to implement changes to the assessment practices that doesn't negatively affect the overall aim of educating "able bodied seafarers" towards the profession.

Although it's impossible to fully control the dynamics of events that unfold during simulations, efforts can be made to manage external factors that may disrupt the testing process. For example, assessors might unintentionally impact the outcomes through their interactions with students during testing, introducing bias or variations (Sellberg et al., 2019). Additionally, in group assessments or scenarios where students interact, there is a risk that students may influence each other's responses, affecting the reliability and consistency of the assessment. In Sellberg (2020) one of the instructors formulated his concerns about consistency in this way, showing an awareness of how slight differences in scenarios and consequently also the tasks create unequal conditions for students:

> I struggle with this issue a lot. I'm spending a lot of time in the simulator and do a lot of examinations and I feel it. It's not fair. I'm not doing a fair assessment. It's so many parameters that affect… just that they are doing a scenario… someone has a small… they don't do the exact same scenarios.

To address these challenges, careful and deliberate design of scenarios for tests of competence is essential. This includes creating standardized conditions that provide every student with an equal opportunity to demonstrate their knowledge and skills. By minimizing variability in the testing environment, such as ensuring that all students face similar tasks and conditions, the reliability of the assessment can be strengthened. This way, assessments can more accurately reflect individual performance and maintain fairness across groups of students.

2.3 Fairness of Simulation-Based Assessments

Fairness in educational assessment emphasizes that examinations should be *equitable* and free from *bias* (Baniasadi et al., 2022). This means that competency tests must provide all students with an equal opportunity to demonstrate their learning, considering diverse backgrounds, personality traits, and potential barriers to access. Defining fairness in assessment is challenging due to the ethical implications that can impact students' learning processes. A major point of debate is the balance between equality and equity, as fair assessment can be viewed through either lens (Murillo & Hidalgo, 2020). Teachers who see fair assessment as a matter of equality generally emphasize qualities such as objectivity and transparency in educational assessment, as highlighted by Murillo and Hidalgo (2020). Objectivity and transparency are ensured through the use of structured assessments, such as written exams which are easy to grade and leave little room for interpretation, allowing clear communication of learning expectations so that students can prepare accordingly. On the other hand, teachers who view fair assessment through the lens of equity argue that measuring student effort should extend beyond grades to recognize individual progress and challenges (Murillo & Hidalgo, 2020). Emphasizing equity means advocating for observing students over time to assess their learning trajectories, believing that understanding and measuring a student's effort requires looking beyond the end product and appreciating the entire learning process (Murillo & Hidalgo, 2020).

In simulation-based tests, fairness is especially critical because such assessments not only measure theoretical knowledge but also practical skills and the ability to engage in professional reasoning and decision-making in action. Therefore, simulation-based tests require careful considerations in regards to equality and equity. On one hand we must ensure that future mariners can reason and act in a competent and professional manner. The equality perspective demands that the criteria for passing the simulation remain consistent and transparent, ensuring that all students are held to the same professional standard. On the other hand, we need to ensure that the assessment process accounts for the diverse backgrounds and learning trajectories of each student. Therefore, an equitable approach would involve adapting simulations to ensuring that all students have the resources and support needed to perform well. For instance, providing additional time or tailored feedback can help address individual needs without compromising the rigor of the test.

Balancing these considerations is crucial in creating an assessment environment where all students are given a fair opportunity to demonstrate their capabilities while maintaining the integrity of the certification process.

2.4 *Bias in Simulation-Based Assessment*

Through interviews with maritime instructors, Sellberg (2020) shows how and why assessments during simulation-based tests of competency can be influenced more by students' conduct and personality traits than their actual understanding of the material. For example, one of the instructors interviewed in Sellberg (2020) explains how he rapidly and automatically starts to assess the students, based on gut feeling:

> Immediately one tries to assess how good the student is. Student 1 is weak. Student 2 is great, Student 3 is kind of weak too, huh? But it can be that this is an effect of how the dialogue unfolds up until the moment we can see here now.

For example, one of the students who delivered an incorrect answer, but in a very confident way, was at first glance assessed as more competent that a student delivering the right answer, but in a hesitant way. It was first after several viewings of the video recorded conversations the group started to reevaluate their first assessments. This makes grading a complex issue since it can be challenging to distinguish between a student's competence and factors like their self-confidence, classroom behavior, or even their personal rapport with the teacher (Sellberg, 2020). This kind of cognitive bias is well known in educational research and occurs when an initial, often ambiguous, assessment of a student influences all subsequent evaluations of that student (Ernstsen & Nazir, 2020). In his PhD thesis on performance assessments on maritime pilot training, Jørgen Ertnsen lists several biases that can occur during simulation-based assessments and have a negative effect on the accuracy and consistency performance assessments in maritime education (Ernstsen, 2020). The different kinds of biases are listed as follows:

Halo Effects and Anchoring Heuristic An assessor's initial, often ambiguous, impression of a student can bias their subsequent evaluations. People often base estimates on an initial value, even if that value is arbitrary. For example, an assessor's initial impression of a student as positive, competent and ambitious might serve as an anchor, biasing their subsequent evaluations to overvalue their performance as we can see in the example from Sellberg (2020).

Recognition-Primed Inferences Assessors may favor actions that are familiar to them, potentially crediting students for solving tasks based on how the assessor would have done it. For example, the student might use to rely more on navigational instruments than visual lookout than preferred by the assessor. However, the student performs a timely, safe and efficient maneuver based on this method and meets the assessment criteria stated in the protocol. While this opens up for interesting discussions and reflection about best practices, one also needs to focus on how well the student met the assessment criteria in order to produce a fair, unbiased, assessment.

Serial Positioning Effects Assessors may have a better memory of actions performed at the beginning and end of a simulation, leading to an inaccurate evaluation of overall performance. For example, one might overemphasize a student's fumbling at the start, interpreting it as a lack of preparedness or skill, or overly focus on a student's strong ability to reflect on their actions, assuming that the trainee demonstrated professional reflection-in-action throughout the exercise.

Availability Heuristic People judge the frequency or probability of an event by how easily they can recall instances of it. This can be influenced by the salience, vividness, or recentness of events, potentially skewing the assessment. For instance, an assessor might overemphasize a recent mistake, even if the student performed well overall.

Illusion of Validity Assessors may feel confident in their judgments even when aware of factors limiting their accuracy. For example, assessors might overestimate their ability to predict a student's future performance based on classroom interactions, grades or performance during simulation.

Representativeness Heuristic People categorize when making inferences and assessments, sometimes ignoring relevant statistical information. For example, an assessor might make assumptions about a student's ability based on their background, potentially leading to an inaccurate evaluation.

Research on bias in educational assessment has primarily focused on disparities faced by women and racial or ethnic minorities, but other underrepresented groups, including LGBTQIA+ individuals, people with disabilities, and those from minority religious backgrounds are also affected by implicit and explicit biases (Ibrahim et al., 2023). Moreover, traditional measures of success, such as standardized test scores, and assessments of competency are increasingly seen as susceptible to bias (Ibrahim et al., 2023). Adapting interventions that have successfully reduced bias in various professional interactions to educational and workplace settings could help address assessment biases according to Ibrahim et al. (2023). These strategies include implicit bias training, aiming to reduce disparities by encouraging individuals to recognize their own biases. However, as Ibrahim et al. (2023) underscores, research in social sciences suggests that merely acknowledging implicit bias is insufficient. Effective training should also incorporate practical strategies to minimize bias, such as *perspective-taking*, *stereotype replacement*, and *counter-stereotype imaging*. These strategies are discussed below:

Perspective-taking is a cognitive strategy that involves putting oneself in the place of another, imagining their thoughts, feelings, and experiences (Ogunyemi, 2021). In the context of assessment, this means envisioning what it might be like to be the person being evaluated, including the challenges they face, their unique strengths, and their individual learning trajectory. This approach is argued to help assessors become more empathetic, reducing the likelihood that their evaluations will be influenced by biases or preconceived notions (Ogunyemi, 2021).

Stereotype replacement is a three-step process involving recognition, reflection and replacement (Ogunyemi, 2021). The steps are outlined as follows:

1. **Recognition:** The first step is to become aware when a reaction or judgment is influenced by an automatic stereotype. This means actively paying attention to any initial thoughts, feelings, or assumptions that might be rooted in bias. For instance, if an immediate reaction aligns with a common stereotype, it's crucial to identify and label that reaction as such.
2. **Reflection:** Once a stereotypical response is recognized, the next step is to pause and reflect on why it arose. This involves questioning the underlying reasons for the automatic thought—considering factors such as past experiences, societal influences, or implicit attitudes. This reflection helps uncover the origin of the biased response, fostering a deeper understanding of one's own thought patterns.
3. **Replacement:** The final step is to consciously replace the stereotypical thought with an unbiased response. This step involves actively challenging the stereotype by focusing on the unique qualities of the individual being evaluated.

This approach can be used to mitigate bias in assessment through self-reflection and active discussions on one's thought process (Ogunyemi, 2021).

Counter stereotype imaging takes a more holistic approach to inclusion and diversity and should be considered as a strategy for the entire educational programme (Ogunyemi, 2021). It involves providing counter-stereotypical examples of maritime expertise, for example to showcase the expertise and contribution to the maritime industry by female or other underrepresented groups. The main idea is that representation and inclusion can play a significant role in reducing implicit bias and fostering a more inclusive educational environment. This approach involves intentionally showcasing individuals who challenge common stereotypes, thus reshaping perceptions and assumptions within the industry. When these professionals share their expertise, experiences, and leadership skills, they serve as powerful counter-narratives to prevailing stereotypes, such as the assumption that leadership roles or technical expertise in the maritime sector are typically held by certain groups.

3 General Guidelines for Strengthening Validity, Reliability, and Fairness of Assessments

Based on the challenges regarding simulation-based assessment and the guidelines that exist regarding valid, reliable, and fair assessment in the pedagogical literature, the following principles for simulator-based competency tests are recommended:

- **Apply predetermined assessment criteria and grading scale.** If you are several assessors, the assessment categories and rating scale should be discussed in

the group so that there is a consensus on *what* it is that is important and *how* different competencies should be assessed.
- **Collaborative assessment** increases both legal certainty and equivalence in examinations. Assessors must strive to be as unbiased as possible and have the professional and pedagogical competence required by university policy and IMO regulations.
- **Document the assessment.** Most simulator providers offer the opportunity to record and save the student's performance during the competency tests. System logs as well as well documented notes are essential for ensuring legal certainty and allow for post-event analysis of cases where the student's self-assessment differs from the assessments made by instructors.
- **Technical breakdowns.** In cases where the simulator technology breaks down so that the assessment test cannot be carried out in a reliable and fair manner, a new examination opportunity needs to be offered to the student within an appropriate time frame. Most higher education institutions have local policies in regard to time frames between examinations and reexaminations to take into consideration, ensuring that all students are given equal opportunities to conduct the competence test.

In sections that follows, we will further outline how to make use of predefined assessment criteria, how you can develop your own assessment model as well as providing guidelines for how to strengthen the assessors' competence.

3.1 Using Predefined Assessment Criteria and Rating Scales

The current assessment system's high demands on valid, reliable, and fair assessment have led to the development of a range of *assessment models* in safety-critical professional training. In safety-critical domains, the focus of assessment models has typically been on so-called *non-technical skills*, i.e. skills described as cognitive and communicative (Flin & O'Connor, 2017). The *technical skills*, i.e., competent handling of various instruments, are, on the other hand, seen as something that is already objectively assessed if routines have been followed and decisions and actions have been carried out in an efficient manner (see e.g. Conceição et al., 2017). The assessment models usually consist of a list of specific behaviors identified as markers of a communicative or cognitive ability, and a scoring scale to measure competence (see Table 1). However, what skills are considered important, as well as what behavior marks a non-technical skill, are specific to the professional practice to be assessed (Flin & O'Connor, 2017). Therefore, there are today a multitude of different assessment models published in the simulator literature across domains. These include, for example, aviation pilot training (Flin et al., 2003), surgeon training (Sharma et al., 2011), anesthesia training (Fletcher et al., 2004) and maritime training (Conceição et al., 2017). In order to implement this kind of model in professional education, the assessment models should therefore meet the learning goals of

Table 1 Behavioral marker system for rating non-technical skills developed by Conceição et al. (2017)

Skill	Behavioral marker
Leadership	Take the initiative. Sets intentions and goals. Establishes and control standards.
Situation awareness	Collects external information. Identifies potential danger or problems. Monitors and reports changes of situations.
Communication	Shares information. Keeps a continuous, clear, and effective flow of information. Promotes a constructive environment for communications.
Teamwork	Considers all the elements of the team. Coordinates the tasks of the team. Assesses the capabilities and corrects procedures.
Decision making	Establishes alternative lines of action. Assesses and verifies consequences of decisions and actions. Considers and shares with others, risks of different lines of action.

the course, and the scoring scale needs to take into account the students' progression. Consequently, the assessment models available in the literature cannot be used directly in different courses but can rather be seen as examples of how non-technical skills can be assess and used as a basis for developing an assessment model based on course specific learning objectives.

The current assessment models place significant emphasis on what should be measured but offer fewer guidelines on how these measurements should be carried out. Two key principles underpin the use of these assessment models.

The first principle is that only *observable behavior* should be assessed. However, there are good reasons to apply this principle with good judgment. There is significant value in exploring how students reason in regards to their navigational decisions during competence tests, as this allows assessors to gain insight into their professional thinking when facing complex and ill-defined navigational challenges (Sellberg et al., 2023). This approach is particularly useful for avoiding Recognition-Primed Inferences—a bias where assessors may favor actions that align with their own problem-solving methods, thereby crediting students based on familiar approaches (Ernstsen & Nazir, 2020).

The second principle is that there should be *clear criteria* of what characterizes the display of good or poor professional behavior (Flin et al., 2003). The main idea is that clear criteria help ensure that assessments of professional behavior are objective and not influenced by subjective opinions or biases. This means that professional standards and best practices need to be identified or established, providing a consistent basis for evaluating actions and behaviors.

Furthermore, Flin and O'Connor (2017) emphasize that contextual factors can negatively influence the assessment. This includes, for example, bias, distractions, or cognitive overload if the situations to be assessed become too complex for the assessor to manage. An important aspect is to calibrate assessment within the assessor group, aiming to increase interrater reliability. In a study on clinical simulation-based assessment, Wong et al. (2024) propose a four-step cycle for finely tuning both the assessors' reasoning about professional competence and the assessment model in use (Fig. 1). In this process, simulation records, such as videos from scenarios, can be revisited to assess students' performance based on predetermined criteria (Buléon et al., 2022). In this way, simulation records can facilitate training for raters, including their calibration and recalibration processes. By allowing raters to evaluate participant performance outside of real-time conditions, videos provide the flexibility for offline assessments and enable external reviews when needed, ensuring a more thorough and objective evaluation (Buléon et al., 2022).

It is of critical importance that assessment models are understood and designed to fit into the teaching practice in which they are to be used (Flin et al., 2003). If a current assessment model is not fit to assess the learning objectives in a course, one might need to develop a new assessment model. How to develop, test, and validate an assessment model is outlined in the next section of the chapter.

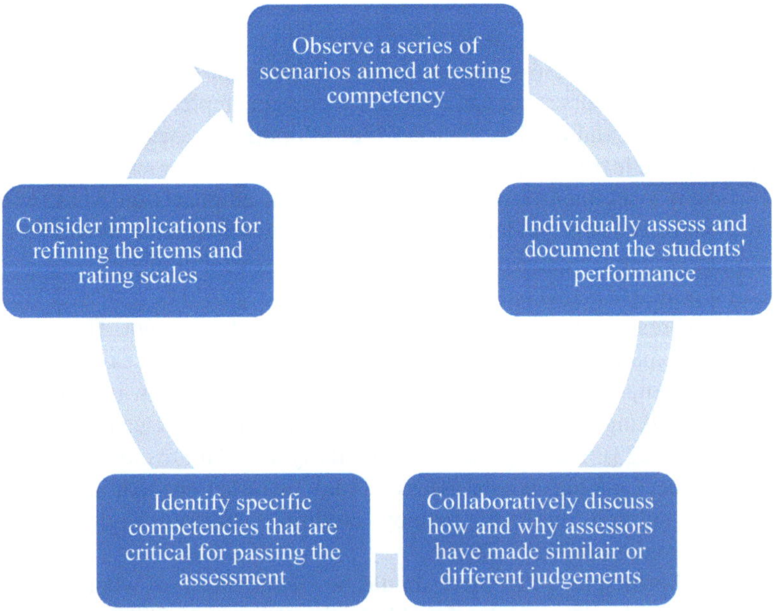

Fig. 1 Key steps involved in strengthening inter-rater reliability (adapted from Wong et al., 2024)

3.2 Developing Assessment Models

In a study on different assessment models in simulation-based training, Hjellvik et al. (2024) emphasizes the importance of robust assessment instruments for enhancing the quality of simulation-based competence tests. Through a systematic literature review, the study synthesizes the findings from 35 research articles to identify common and particular characteristics of assessment instruments in different professional domains where simulation-based training and assessment are important parts of education. As a result of this analysis, Hjellvik et al. (2024) develops an empirically grounded five step method for the development and validation of assessment instruments to be used for simulation-based tests of competency. In this chapter, we have adapted their five-step model towards practical use in simulation-based training. In our three-step model, the process steps are mapped towards best practices in higher education, putting focus on keeping a close link between learning objectives and assessment items (Table 2). We also left out matters involved in designing the scenario from the process steps since matters of scenario design is covered in chapter "Preparing for Simulation Through Briefing" of this book.

3.3 Strengthening the Assessor's Competence

An important point made in the literature is that the quality of the assessment depends on the assessor's competence (Sellberg, 2020). Carrying out assessments that both examine courses in higher education and which at the same time form the basis for issuing certificates involves carrying out assessments where values such as validity, reliability and fairness are put at the forefront and come into conflict with the higher goals of education, i.e. to educate competent mariners for a professional life at sea. At first glance, it may seem that the professional competence of experienced sea captains should be able to guarantee a valid assessment of competences for maritime certificates and examinations in a navigation course. The professional knowledge is an absolute prerequisite for being able to assess the competence of maritime students, but it is also of importance that maritime instructors are given the chance to develop a professionalized assessment practice tailored to the specific needs of assessing maritime skills and competencies. Moreover, when the aim is to assess both technical and non-technical skills, assessments do not only rely on the assessors' professional experience and competence, but also on whether the assessor has a theoretical understanding of the non-technical skills and how they can be measured in order for the assessment to be carried out in a competent and consistent manner (e.g., Sharma et al., 2011).

To strengthen the assessors' competence Wong et al. (2024) recommend the following activities:

Table 2 Three steps to creating and validating an assessment model (adapted from Hjellvik et al., 2024)

Process step	Activities involved
1: Identify learning objectives, available metrics and appropriate items	**1. Identify Learning Objectives**: Learning objectives are the specific skills, knowledge, or abilities you want the test to measure. These should align with what you want learners to achieve by the end of their training or education. **2. Determine Which Objectives to Assess**: Not every learning objective needs to be assessed through a test, so you need to decide which are most important for the competence test. **3. Formulate Appropriate Items:** For each learning objective, you need to create specific questions or tasks (items) that will assess that objective. These items should directly measure the learner's understanding or skill level related to each objective. **4. Identify Sources of Information**: Different types of information sources can be used to assess different skills or knowledge. You need to decide the best way to measure each item based on what provides the most accurate reflection of the learner's competence.
2: Pilot test items and rating scale	**1. Establish Content Validity:** This means ensuring that the metrics or indicators you include truly represent the skill, ability, or construct you're trying to measure. **2. Including Metrics with Discriminatory Ability**: This refers to choosing metrics that can effectively distinguish between high and low performers. If a metric can't differentiate between different levels of performance, it may not be useful for evaluation. **3. Excluding Metrics with High Inter-Item Correlation**: Metrics that have high inter-item correlation are very similar to one another, meaning they measure almost the same thing. Including too many similar metrics can be redundant and can make your assessment less efficient.
3: Update and validate assessment model	**1. Update assessment model:** This involves making adjustments to the overall assessment framework based on the feedback and data collected during a previous phase. It might include revising items, adjusting rating scales, or modifying the overall structure of the assessment. **2. Pilot test new items:** After making significant changes to the assessment model, it's important to run a small-scale pilot test. This helps you understand how well the new or revised items work before implementing them in a larger assessment. **3. Expert validation of assessment model:** Expert validation helps ensure the content is accurate, relevant, and aligned with industry standards or best practices, providing feedback on whether the items, rating scales, and overall approach accurately measure the desired skills or knowledge areas.

- **Strengthen partnerships with learners:** Engage assessors in dialogic feedback with students, utilizing both formal course evaluations and informal discussions. This collaborative approach fosters a supportive learning environment.
- **Promote peer mentoring and observation:** Implement activities that encourage peer mentoring and observation/auscultation. These practices allow both new and experienced assessors to learn from one another and to justify their judgments, thereby enhancing the quality of assessment processes.

- **Facilitate knowledge exchange among assessors:** Develop a mutual understanding of assessment terminology through workshops, seminars and conferences. This clarity helps assessors define the expected levels of achievement and align their definitions of the competencies being assessed within each item of the assessment model.
- **Leverage the energy of "assessment champions"**: As described by Wong et al. (2024), assessment champions play a crucial role in driving improvements in assessment practices. By leading initiatives and generating momentum, these champions can foster innovative approaches to assessment.
- **Build, maintain and/or contribute to a digital assessment hub**: Establish a digital platform to share information about assessment resources, key considerations, and digital exemplars. For instance, the International Association of Maritime Universities (IAMU) is developing a digital assessment hub consisting of validated scenarios, assessment models and rating scales to be used by members of IAMU.

In chapter "Instructors as Co-Researchers: Strengthening Professional Practice Through Collaborative Research", the professional development of maritime instructors and assessors is discussed further.

References

Baniasadi, A., Salehi, K., Khodaie, E., Bagheri Noaparast, K., & Izanloo, B. (2022). Fairness in classroom assessment: A systematic review. *The Asia-Pacific Education Researcher*, 1–19.

Brennan, R. L. (2023). *Educational measurement*. Rowman & Littlefield.

Buléon, C., Mattatia, L., Minehart, R. D., Rudolph, J. W., Lois, F. J., Guillouet, E., et al. (2022). Simulation-based summative assessment in healthcare: An overview of key principles for practice. *Advances in Simulation, 7*(1), 42.

Conceição, V. P., Basso, J. C., Lopes, C. F., & Dahlman, J. (2017). Development of a behavioral marker system for rating cadet's non-technical skills. *TransNav: International Journal on Marine Navigation and Safety of Sea Transportation*, 11.

Conn, C. A., Bohan, K. J., Pieper, S. L., & Musumeci, M. (2020). Validity inquiry process: Practical guidance for examining performance assessments and building a validity argument. *Studies in Educational Evaluation, 65*, 100843.

Emad, G., & Roth, W. M. (2008). Contradictions in the practices of training for and assessment of competency: A case study from the maritime domain. *Education+Training, 50*(3), 260–272.

Ernstsen, J. (2020). *Reducing the subjective impact in maritime simulator assessment A performance assessment tool for maritime pilotage operations*. Doctoral dissertation from University of South Eastern Norway.

Ernstsen, J., & Nazir, S. (2020). Performance assessment in full-scale simulators–A case of maritime pilotage operations. *Safety Science, 129*, 104775.

Fletcher, G., Flin, R., McGeorge, P., Glavin, R., Maran, N., & Patey, R. (2004). Rating non-technical skills: Developing a behavioral marker system for use in anaesthesia. *Cognition, Technology & Work, 6*(3), 165–171.

Flin, R., & O'Connor, P. (2017). *Safety at the sharp end: A guide to non-technical skills*. CRC Press.

Flin, R. H., Martin, L., Goeters, K. M., Hormann, H. J., Amalberti, R., Valot, C., & Nijhuis, H. (2003). Development of the NOTECHS (non-technical skills) system for assessing pilots' CRM skills. *Human Factors and Aerospace Safety, 3*(2), 97–119.

Gekara, V. O., Bloor, M., & Sampson, H. (2011). Computer-based assessment in safety-critical industries: The case of shipping. *Journal of Vocational Education & Training, 63*(1), 87–100.

Hjellvik, S., Mallam, S., Giskeødegård, M. F., & Nazir, S. (2024). Review on competency Assessment instrumentation in computer-based simulation. *Technology, Knowledge and Learning*, 1–30.

Ibrahim, H., Juve, A. M., Amin, A., Railey, K., & Andolsek, K. M. (2023). Expanding the study of bias in medical education assessment. *Journal of Graduate Medical Education, 15*(6), 623–626.

International Maritime Organization. (1978). *International Convention on Education, Training, Certification and Watchkeeping for Seafarers 1978, as amended (Manilla Amendments)* STCW Convention and STCW Code, London, (2011).

Kang, H., Thompson, J., & Windschitl, M. (2014). Creating opportunities for students to show what they know: The role of scaffolding in assessment tasks. *Science Education, 98*(4), 674–704.

Mirza, N., Cinel, J., Noyes, H., McKenzie, W., Burgess, K., Blackstock, S., & Sanderson, D. (2020). Simulated patient scenario development: A methodological review of validity and reliability reporting. *Nurse Education Today, 85*, 104222.

Murillo, F. J., & Hidalgo, N. (2020). Fair student assessment: A phenomenographic study on teachers' conceptions. *Studies in Educational Evaluation, 65*, 100860.

Ogunyemi, D. (2021). A practical approach to implicit bias training. *Journal of Graduate Medical Education, 13*(4), 583–584.

Reid, J., Stone, K., Brown, J., Caglar, D., Kobayashi, A., Lewis-Newby, M., et al. (2012). The simulation team assessment tool (STAT): Development, reliability and validation. *Resuscitation, 83*(7), 879–886.

Sellberg, C. (2020). Pedagogical dilemmas in dynamic assessment situations: Perspectives on video data from simulator-based competence tests. *WMU Journal of Maritime Affairs, 19*(4), 493–508.

Sellberg, C., Lindmark, O., & Lundin, M. (2019). Certifying navigational skills: A video-based study on assessments in simulated environments. *TransNav, International Journal on Marine Navigation and Safety of Sea Transportation, 13*(4).

Sellberg, C., & Lundin, M. (2018). Tasks and instructions on the simulated bridge: Discourses of temporality in maritime training. *Discourse Studies, 20*(2), 289–305.

Sellberg, C., Lundin, M., & Säljö, R. (2022). Assessment in the zone of proximal development: Simulator-based competence tests and the dynamic evaluation of knowledge-in-action. *Classroom Discourse, 13*(4), 365–385.

Sellberg, C., Wiig, A. C., & Säljö, R. (2023). Mastering the artful practice of navigation: The situated endorsement of professional competence in post-simulation evaluations. *Studies in Educational Evaluation, 72*, 101–111.

Sharma, B., Mishra, A., Aggarwal, R., & Grantcharov, T. P. (2011). Non- technical skills assessment in surgery. *Surgical oncology, 20*(3), 169–177.

Wong, W. Y. A., Gormley, G. J., Haughey, S., Chong, S. W., & Wilson, C. B. (2024). Better together: An assessor support roadmap. *The Clinical Teacher, 21*(1).

Charlott Sellberg is an Associate Professor in IT and Learning at the Department of Applied IT, University of Gothenburg. With a multidisciplinary background in Cognitive Science and Human-Computer Interaction, she earned a PhD in Education from the University of Gothenburg in 2018. Her doctoral work examined both the theoretical and practical aspects of simulation-based training in the maritime industry. Since then, she has continued to study the role of simulations in maritime training and assessment, in areas such as basic safety training, SAR operations, and advanced ship handling, continuing to deepen the understanding of simulation as a learning environment and informing educational practices within the maritime domain.

Instructors as Co-Researchers: Strengthening Professional Practice Through Collaborative Research

Mona Lundin, Annika Lantz-Andersson, Ewa Skantz, and Pia Williams

Abstract This chapter explores the evolving landscape of collaborative research between instructors and researchers, emphasizing the need for practice-centered approaches to strengthen educational practices and support professional development. Grounded in traditions such as Action Research, Participatory Action Research, and Design-Based Research, the chapter highlights how teachers can serve not only as recipients of research but as co-researchers and co-producers of knowledge. It outlines the historical and theoretical roots of these approaches, including models like Learning Study, Teacher Development Partnership Research, and Research-Practice Partnerships, while acknowledging their potential to bridge the persistent gap between theory and practice. The chapter also critically examines the tensions and challenges inherent in these collaborations. Issues of power asymmetry, role ambiguity, trust uncertainty, and institutional constraints often limit meaningful participation and equitable knowledge production. The concept of the Third Space is introduced as a framework to support more balanced, dialogic engagement between academic and practitioner knowledge expertise. Ultimately, the chapter calls for a nuanced understanding of collaborative educational research that recognizes its political, cultural, and structural dimensions, and foregrounds sustained, reciprocal relationships as key to a more professionalized instructional practice.

M. Lundin (✉) · A. Lantz-Andersson · E. Skantz · P. Williams
Department of Education, Communication and Learning, University of Gothenburg, Gothenburg, Sweden
e-mail: mona.lundin@ped.gu.se; annika.lantz-andersson@ped.gu.se; ewa.skantz@gu.se; pia.williams@ped.gu.se

C. Sellberg (ed.), *Simulations in Maritime Education and Training*, WMU Studies in Maritime Affairs 12, https://doi.org/10.1007/978-3-032-08394-4_7

1 Introduction

Over the years, numerous initiatives, approaches, and models for collaboration between researchers and practitioners[1] have emerged and been implemented to improve the quality of educational research and align it more closely with the needs of instructional practice (Zeichner & Noffke, 2001). As interest in practice-centered research continues to grow, these collaborative models are increasingly referred to by various terms, reflecting diverse theoretical and methodological perspectives. Practice-centered approaches are found in various research traditions, including Action Research (Reason & Bradbury, 2001), Participatory Research (Cornwall & Jewkes, 1995), or Participatory Action Research (Kemmis & McTaggart, 2005), and more recently, Critical Participatory Action Research (Kemmis et al., 2014). While these approaches sometimes overlap in how they describe the collaborative process, they originate from distinct traditions. What unites them is a shared goal: to deepen the understanding of and improve local instructional practices. This reflects a dual focus of improving the quality of educational research while aligning it more closely with the practical needs of the field (Zeichner & Noffke, 2001). The literature also highlights the complexities of research collaborations, including challenges such as gaining access to instructional practices, building equitable instructor-researcher relationships, addressing differing competencies and structural constraints (e.g., time), and ensuring that the research remains relevant to real-world instructional challenges (e.g., Nutley et al., 2008).

In Sweden, practice-centered research has increasingly been emphasized as both an objective and a means for strengthening the scientific foundation of instructional practice. In recent years, the discussion surrounding teachers' participation in practice-centered research has gained momentum as a series of governmental initiatives have been issued to enhance the collaboration between research and school practices (Magnusson & Malmström, 2023). The most recent example is the 2024 government inquiry into reforming teacher and preschool teacher education (SOU, 2024, p. 81), which highlights the importance of integrating research and practice. This inquiry proposes making a previous initiative permanent—namely, the ULF agreement from 2017 *(in Swedish: Utveckling, Lärande, Forskning)*. The initiative reflects a strong national commitment to building "A school based on scientific evidence" (*in Swedish: En skola på vetenskaplig grund*). Its primary aim is to develop and test sustainable models for long-term collaboration between researchers and practitioners, including teachers, encompassing research, educational development, and teacher education. Another key initiative is the establishment of the Swedish Institute for Educational Research (*in Swedish: Skolforskningsinstitutet*), whose mandate is to support instructors in planning, implementing, and evaluating

[1]Throughout this chapter, the terms practitioners, instructors, educators and teachers are used interchangeably to refer to the counterparts of researchers in research-practice collaborations, to include compulsory school teachers, higher education educators, as well as instructors involved in professional development efforts.

teaching based on research-informed methods. The Institute allocates funding specifically for practice-based educational research (*in Swedish: praktiknära forskning*; sometimes referred to as "practice-near school research" in policy analysis) with the explicit aim of developing models rooted in the practical challenges, needs, and questions encountered in educational practice. Ultimately, these initiatives seek to contribute to improved learning and development outcomes for children and students. In parallel, the Commission of Inquiry on Practice-Based School Research (SOU, 2018, p. 19) proposed strategies for fostering long-term collaboration in practice-based educational research between teacher education institutions and school leadership. These initiatives share a common assumption: that research-informed teaching, grounded in proven experience, enhances educational quality, goal achievement, and student learning. Embedded in the political agenda, they reflect high expectations for the potential of such collaborative research to drive meaningful educational improvement.

It is often assumed that aligning research collaboration directly with the immediate needs of local instructional practice effectively bridges the long-standing gap between research and practice. While such alignment can increase relevance and engagement, this perspective oversimplifies the complex dynamics of collaborative knowledge production. Another widely held belief is that practice-centered research, in which instructors are both consumers and co-producers of knowledge, naturally leads to school development[2] and ensures that improvements are grounded in scientific evidence. Although this belief holds great potential, it presupposes ideal conditions for participation and overlooks the tensions that may arise when merging professional teaching experience with academic research standards. An almost equally common notion is that providing teachers with time to participate in research collaborations is sufficient to foster meaningful school development. Practice-based research, as argued by Hardy et al. (2017, p. 424) is closely tied to embodied understandings of professional learning that are "personal, professional, and political." For research-practice collaborations to be stable and productive, teachers must be supported with the necessary conditions for meaningful participation. Reciprocity plays a central role in amplifying diverse perspectives and fostering a culture of mutual exchange. Key factors for successful collaboration include shared goals, aligned values, and a balance of authority and contributions among all participants (Coburn et al., 2013). However, this raises important questions about the structural and cultural challenges instructors face, such as workload, institutional support, and competing demands. Under current conditions, the scope for change may be more limited than often acknowledged, calling for a more nuanced understanding of what such collaborations can realistically achieve. This chapter aims to discuss some of the current approaches to researchers-practitioners collaborations, focussing on the scope for change and the challenges for participation.

---

[2] School development also includes advancements in contexts such as higher education and lifelong learning.

2 Collaborative Pathways: Empowering Teachers as Co-Researchers

As early as Dewey (1929), scholars argued that teachers must be active participants in the research process for its outcomes to meaningfully inform teaching. Since then, educational research has emphasized teacher involvement as a means for professional development and strengthening the knowledge base for teaching (see e.g., Cole & Knowles, 1993). While earlier efforts often put teachers in the role of data collectors (Zeichner & Noffke, 2001), contemporary approaches promote collaboration as a dual benefit: supporting professional learning and improving instructional practice, while helping researchers align their work with classroom realities.

Action Research, an umbrella term for inquiry-driven change (Reason & Bradbury, 2001), was first coined by Kurt Lewin in the 1940s. A fundamental assumption is that change necessitates a continuous research process, responsive to teachers' needs. Lewin (1946) described Action Research as a circle with four steps: plan, act, observe, and reflect. The model later developed into a spiral, representing an ongoing process in which reflection generates new plans and any step can restart the cycle. When the model was taken up in education in the U.K. in the 1970s, the active role of teachers as partners in developing and implementing new curriculum came to the fore, notably in the work of Lawrence Stenhouse and John Elliot. Stenhouse (1975) argued that teachers needed to study their own work, coining the term *Teachers-As-Researchers* to highlight teachers' active involvement in research. Supported by universities, the movement gained momentum and led to the creation of the Classroom Action Research Network (later Collaborative Action Research Network, ARN) and to the journal Educational Action Research, (Somekh, 2009; Zeichner & Noffke, 2001).

Approaches emphasizing teachers as researchers in school development also include *Participatory Research* (PR) (Lundberg & Starrin, 2001), *Participatory Action Research* (PAR) (Brydon-Miller & Maguire, 2009), and *Community-based Participatory Research* (CBPR) (Jull et al., 2017). These approaches emerged around the same time in response to traditional academic research, which often set the agenda of inquiry (Brydon-Miller & Maguire, 2009). Rooted in critical social theory and Paulo Freire's pedagogy of empowerment, a central tenet is the "transformative social justice agenda," addressing issues of inequality and power, often raised by marginalized groups (Brydon-Miller & Maguire, 2009). Thus, PAR is inherently political and should be "conducted in a manner that ordinary people can participate in the whole process from the formulation of the issues to the analysis of the results and the discussion of how they should be used" (Lundberg & Starrin, 2001, p. 16). While researchers might intervene to mobilize and organize investigation and action, the core idea is that practitioners take an active role rather than being passive observers. The goal is to build so-called alliances between researchers and practitioners in planning, implementing, and disseminating the research (McIntyre, 2008). A key argument for participant involvement throughout the process is their valuable context-specific knowledge, which might otherwise be overlooked by researchers. Consequently, knowledge developed through such collaboration is thought to become more profound, enriched, and social when produced collectively (Lundberg & Starrin, 2001).

In Sweden, PAR is often taken as synonymous to *Learning Study,* a modified version of Japan's Lesson study (Marton & Runesson, 2014). Learning Study aims to improve teaching methods and student learning, and typically starts in teachers' own didactical challenges. This method involves cyclical classroom interventions focused on subject-specific learning objects, based on phenomenography, variation theory, and evidence from practice. Researchers introduce variation theory principles, relevant literature, and collaborate with teachers on lesson design, implementation, and evaluation. In the cyclical process, dialogues among teachers as well as between researchers and teachers are essential. To support improvement and change, collaboration relies on the unique knowledge and experience of each professional group: researchers bring theory and methods and teachers offer practical, context-specific expertise (Carlgren, 2012). Important factors that help this kind of collaboration work well include having a clear structure, defined responsibilities, and clear division of tasks (Lillejord & Børte, 2016). One conclusion drawn is that a precondition for successful partnerships is an open and trusting relationship between partners. However, several questions remain about the long-term implications of such collaborations for teachers and in instructional practices, as well as whether they genuinely enable teachers to be equal co-producers in, and contributors to research (Stetsenko, 2017).

In the 1980s, a branch of action research called *Practitioner Research* was developed in the US, which focused on tools for data collection and sharing teachers' findings, such as journal writing and observation. Results were published in subject-specific journals, and conferences were held where researchers and teachers shared knowledge (Zeichner & Noffke, 2001). *Collaborative Action Research*, as argued by Carr and Kemmis (1986), marked a shift in the scientific field toward new research approaches. They applied Habermas's theory of knowledge to distinguish three types of action research:

- *Technical action research* focusing on applied skills to achieve objectives.
- *Practical action research* centering on reflective practice and qualitative methods to help teachers examine their practices.
- *Critical action research* addressing the social and political dimensions of education, linking individual practice to broader institutional structures.

Critical action research aims to improve practice, deepen understanding, and address the social context of education, highlighting its political aspects (Carr & Kemmis, 1986). However, over time, action research became increasingly institutionalized and detached from its emancipatory roots, functioning more as a method aligned with orthodox research paradigms (Carr & Kemmis, 2005). Across these above-mentioned traditions, reflective dialogue tend to be prioritized over data production, with the goal of emancipating both teachers and researchers (Carr & Kemmis, 1986; Zeichner & Noffke, 2001). Building on these historical and critical strands, current work on collaboration between researchers and practitioners includes a wide array of partnership-oriented approaches. In what follows, we highlight four central examples: Teacher Development Partnership Research (TDPR), Research-Practice Partnerships (RPPs), Design-Based Research (DBR), and a Third Space that mediates academic and practitioner knowledge.

Teacher Development Partnership Research (TDPR) is based on a hermeneutic perspective (Ryle, 1949) and emphasizes the interpretative relationship between researchers and teachers, focusing on their collaborative dynamic (Cole & Knowles, 1993). The TDPR process typically unfolds in stages, from initiation to reporting, and is guided by specific questions related to:

(1) *Technical issues:* Methods for generating, managing, and organizing data.
(2) *Personnel issues:* Defining and evolving roles and responsibilities throughout the process.
(3) *Procedural issues:* Setting a clear time frame for the research.
(4) *Ethical issues:* Addressing confidentiality, differing interpretations, and ensuring equity in influence.
(5) *Political issues:* Determining who will use the research and for what purposes.
(6) *Educational issues:* Evaluating how participation impacts practice and supports professional development.

In *Teacher Development Partnership Research*, collaborative classroom research is described as a cyclical feedback process that extends throughout the field work (Cole, 1989), and involves sharing and developing ideas as well as discussing specific instructional activities. For researchers, this means further theoretical examination, whereas for teachers it involves scientific exploration of practice, leading to professional development for both groups.

Research-Practice Partnerships (RPPs), introduced by Coburn et al. (2013), are a practice-centered approach, created in response to the pressure on schools to implement evidence-based practices and the lack of support teachers face in applying research that often does not address their immediate needs in teaching. Existing research is often seen as irrelevant or arrives too late to impact instructional practice. In response, RPPs focus on issues directly relevant to teachers, emphasizing long-term, collaborative partnerships between practitioners and researchers to solve problems and improve outcomes (Coburn et al., 2013). Teachers are viewed as co-designers and equal participants in the research process, with research questions initiated by local instructional needs (Coburn & Penuel, 2016).

Design-Based Research (DBR), rooted in pragmatic philosophy and Dewey's applied theory of practice, is not a single approach but a family of strategies aimed at developing theories, artifacts, and practices that impact teaching and learning in real-world settings (Barab & Squire, 2004). Key features of DBR include theory development, intervention through design, work in naturalistic contexts, and iterative processes of refinement (Cobb et al., 2003). High-quality DBR studies focus on authentic educational contexts and significant interventions (Anderson & Shattuck, 2012). Grounded in both theory and context, DBR produces theoretical constructs that extend beyond specific settings and seeks to generate evidence-based claims that advance contemporary theoretical knowledge (Barab & Squire, 2004).

Third Space offers an approach for bridging the gap between theory and practice in teacher education and research-practice collaborations (e.g., Arhar et al., 2013; Zeichner, 2010). It promotes more equitable partnerships in which participants (teachers, students, and researchers), engage on equal terms, each contributing their own knowledge and expertise. As a hybrid space, Third Space integrates academic and practical knowledge, rejecting binaries such as theory versus practice or practitioner versus researcher (Zeichner, 2010). It supports the co-construction of researchable questions grounded in practice, fostering mutual learning and sustainable collaboration. Importantly, it can provide neutral grounds for participants to negotiate meaning, step outside their comfort zones, and generate shared insights (Lillejord & Børte, 2016). In this way, Third Space functions both as a tool for change and a platform for developing innovative networks that improve educational practice and inform policy (Zeichner, 2007, 2010).

3 Navigating the Complexities of Researcher–Practitioner Collaboration

While numerous studies emphasize the importance of involving practitioners as co-producers in educational development, these collaborative approaches face significant challenges, particularly in achieving meaningful participation. Within the *Participatory Action Research* (PAR) tradition, for instance, the very notion of participation is open to interpretation. Cornwall and Jewkes (1995) propose that participation should not be viewed as static but rather as existing along a continuum. They outline four distinct models of participation that define the relationship between researchers and practitioners:

- *Contractual*—participants are recruited by researchers to contribute to pre-designed inquiries or experiments.
- *Consultative*—participants are consulted for their opinions prior to the implementation of research decisions.
- *Collaborative*—researchers and participants work together, although the project is still primarily initiated and directed by researchers.
- *Collegiate*—researchers and participants collaborate as equals, bringing different skills to a process of mutual learning and shared control over research.

The choice of participation model is often influenced by researchers' epistemological beliefs and their willingness to share authority with partners outside academia (Cornwall & Jewkes, 1995). These decisions inevitably affect not only the research process and outcomes but also the degree to which participants are empowered to enact change. Within education, PAR is increasingly adopted, yet concerns remain. Brydon-Miller and Maguire (2009) caution that PAR risks becoming a depoliticized, problem-solving tool for teachers, thus diluting its foundational commitment to social justice. Nevertheless, they acknowledge that PAR frameworks can still empower teachers to uncover and address systemic inequities in schools and to develop more inclusive, democratic teaching practices.

In the context of *Teacher Development Partnership Research* (TDPR), the researcher-teacher relationship is central. Here, both parties are seen to bring distinct competencies, engaging as joint theorists in a shared process of meaning-making. Cole and Knowles (1993) argue that this collaborative inquiry should foster professional development for both researchers and teachers. When designed collaboratively, research becomes a communicative process that enhances mutual understanding and supports teachers' analytical growth and disciplinary foundations, i.e., increased professional understanding (Cole, 1989). The relationship is thus seen as symbiotic, interdependent, and mutually beneficial. Despite these ideals, TDPR frequently reverts to traditional research dynamics where researchers have primary control, and teachers have limited engagement. This gap in collaboration stems from a lack of ongoing negotiation around roles, responsibilities, and commitment. Cole and Knowles (1993) describe this as "cooperation," rather than collaboration, which often leads to role confusion and limited mutual, collaborative interpretation of practice.

Research Practice Partnerships (RPPs) similarly highlight tensions. Researchers and practitioners often operate within different cultures and timelines. While teachers look for immediate solutions to here-and-now classroom dilemmas, researchers has the time to proceed cautiously, grounded in evidence (Coburn et al., 2013). Language barriers further complicate collaboration, as parties may lack a shared vocabulary for discussing practice (Coburn & Penuel, 2016). The line between researchers and practitioners can be blurred and the roles may be vague and unclear. Farrell et al. (2019) emphasize the need to continually revisit and clarify roles and contributions throughout the a project to prevent stagnation or confusion.

The Third Space approach foregrounds deeper structural challenges, chiefly the divide between academic and practical knowledge, and power asymmetries between universities and e.g., schools. Achieving an equal partnership that involves shared discourse is a complex task. Klein et al. (2013) emphasize the importance of redefining roles and developing mediating practices that integrate diverse perspectives. Williams (2014) shows how working within a third space can help teacher educators rethink professional identities and foster reflective practice. Arhar et al. (2013) describe the success of an apprentice-style model that valued both teacher and university expertise equally, showing the potential of a third space to support mutual commitment and trust.

Within *Design-Based Research* (DBR), the researcher-practitioner partnership is again a key concern. Researchers in DBR are not mere observers but are actively shaping the learning environment, they are engineering the very same activities they are making claims about (Barab & Squire, 2004). Amiel and Reeves (2008) caution against researchers acting as external consultants and stress the need for shared ownership over problem identification and design. Dede (2005) notes that tensions may arise between practitioner-identified issues and research-driven agendas, but such tensions can also be productive sites for dialogue. These partnerships must navigate researchers' lack of contextual knowledge and teachers' limited time or training for rigorous research. As Anderson and Shattuck (2012) describe, the DBR process requires continuous negotiation from problem definition to implementation and dissemination.

A main challenge across these approaches is trust (Coburn et al., 2013). Past experiences of research intervention and broken promises might have fostered scepticism among practitioners. Furthermore, balancing classroom relevance with broader scalability of research findings remains contentious. Bevan et al. (2018) suggest taking generalizability as the practical uptake of research findings across diverse settings. Similarly, Carlgren (2012) argues that theoretically grounded analysis, can yield generalizable insights even in context-bound studies, such as Learning Study, allowing for broader inferences despite their situated nature.

4 The Politics of Participation: Navigating Tensions in Collaborative Research

A range of challenges complicates the pursuit of equal participation in collaborative educational research. As mentioned previously, one of the most foundational issues lies in the ambiguous and contested nature of participation itself. As Cornwall and Jewkes (1995) point out, participation should not be understood as a fixed category, but rather as a continuum ranging from minimal engagement to full co-ownership of the research process. However, in practice, achieving the more egalitarian forms, such as collaborative participation, often proves difficult. Researchers may claim participatory intentions, yet maintain control over core decisions, subtly reinforcing traditional hierarchies.

This issue of control is closely tied to deeper power imbalances within researcher-practitioner collaborations. Despite the rhetoric of partnership, many projects still reflect an asymmetrical structure in which researchers lead, while teachers contribute only marginally. This is particularly evident in Teacher Development Partnership Research (TDPR) and Participatory Action Research (PAR), where the ideal of co-production is often undermined by a lack of clarity around roles and insufficient negotiation of responsibilities (Cole & Knowles, 1993). Without explicit dialogue about who does what, and why, collaboration risks becoming symbolic rather than substantive. Some scholars warn of a growing depoliticization as PAR becomes more normalized in institutional settings. To reiterate a previous point, if PAR is reduced to a practical tool for classroom problem-solving, its critical, emancipatory roots may be weakened. Brydon-Miller and Maguire (2009) caution that unless power dynamics and social inequalities remain central concerns, PAR runs the risk of reinforcing the status quo rather than challenging it. This tension between immediate pragmatic gains and long-term structural change continues to pose a dilemma for researchers and practitioners alike.

Another persistent challenge relates to the temporal and cultural tension between researchers and practitioners. Teachers work under intense time pressures and policy demands, often needing rapid solutions to pressing classroom problems. Researchers, in contrast, are typically bound by the slower pace of academic inquiry and often hesitate to act without a strong empirical foundation. These divergent rhythms can hinder collaboration, especially when the urgency of practice clashes with the deliberateness of research (Coburn et al., 2013). Additionally, communication barriers further complicate these partnerships. A shared vocabulary for analyzing and discussing practice is frequently lacking, leading to misunderstandings and missed opportunities for collaborative meaning-making. As Coburn et al. (2016) note, even the most well-intentioned collaborations may stumble if researchers and practitioners cannot speak to each other in conceptually aligned ways.

The theory-practice divide presents another profound difficulty, particularly for Third Space approaches and Research Practice Partnerships (RPP). These models aim to bridge academic and practitioner knowledge, yet power asymmetries and differing epistemologies can stand in the way. Simply bringing together stakeholders from different spheres is not enough; mediating spaces and practices are required to foster trust, shared understanding, and mutual learning. Without such scaffolding, partnerships risk becoming merely symbolic and short-term.

As already mentioned, trust indeed is a recurring theme across the literature and an essential precondition for meaningful participation. Many teachers have had prior negative experiences with research participation, such as being treated as data sources rather than co-creators, or have encountered projects that made promises but failed to deliver or maintain engagement. These experiences generate understandable skepticism, making it even more challenging to establish genuine partnerships based on reciprocity and mutual respect (Coburn et al., 2013).

Another tension involves generalizability. Collaborative research is often deeply rooted in local contexts, raising doubts about how findings might be applied elsewhere. While some, like Bevan et al. (2018), argue for a more practice-oriented notion of generalizability, one based on adaptation rather than replication, this view is still contested, particularly within academic institutions that prioritize large-scale, transferable results. Similarly, approaches like Learning Study, which are grounded in situated, context-specific analysis, continue to face critique for their limited scalability, despite efforts to identify transferable patterns through theoretical grounding (Carlgren, 2012).

Finally, the issue of researchers' dual roles, as both facilitators and investigators, is especially discussed in Design-Based Research (DBR). Researchers in DBR are not neutral observers but active designers of educational interventions, which raises questions about the objectivity and validity of their claims (Barab & Squire, 2004). Moreover, while DBR promotes practitioner voice in the design processes, the balance of power can still lean toward researchers, particularly if the school-identified problems are overshadowed by research agendas or methodological concerns.

Taken together, these challenges suggest that participation in educational research is not simply a methodological choice, but a deeply political and epistemological commitment. It requires continual reflection, negotiation, and humility on the part of researchers, especially careful attention to whose knowledge counts, how decisions are made, and what collaboration entails in practice. Sustainable partnerships benefit from shared agenda setting, agreed decision processes, and reciprocity in the use of results. They also depend on institutional recognition of practitioner inquiry as legitimate knowledge production and on resources that provide time for iterative cycles of design, implementation, and reflection. This stance resonates with Stetsenko's view of learning and development as transformative, collective activities directed toward changing both practitioners and the educational worlds they inhabit (Stetsenko, 2017). When such conditions are in place, collaborations are better positioned to generate knowledge that is both locally meaningful and theoretically generative.

Key Considerations for Instructors in Research Collaborations
Clarify Your Role Early

- Define your role and responsibilities clearly from the start, are you a co-designer, data provider, or co-author? This prevents misunderstandings later.Negotiate Goals and Expectations
- Ensure that both your professional priorities and the researchers' objectives are aligned, or at least mutually acknowledged.Bring Your Expertise
- Your classroom experience and contextual knowledge are crucial, do not undervalue your voice or defer automatically to academic knowledge.Ask About the Research Design
- Understand the methods, timelines, and expected outcomes. Ask how the research will benefit your practice and your students.Expect Ongoing Dialogue
- Collaboration is not a one-time input, look for partnerships that include regular reflection, feedback, and shared learning.Consider Time and Workload
- Be realistic about your availability. Will participation be supported by your leadership or built into your schedule?Stay Critical and Reflective
- Engage critically, question assumptions, share concerns, and suggest adaptations to fit your context.Think Long-Term
- How will this collaboration contribute to your professional development, school improvement, or sustained instructional change?Discuss Ownership and Credit
- Talk openly about authorship, presentation of findings, and how results will be shared or implemented.

References

Amiel, T., & Reeves, T. C. (2008). Design-based research and educational technology: Rethinking technology and the research agenda. *Journal of Educational Technology & Society, 11*(4), 29–40.

Anderson, T., & Shattuck, J. (2012). Design-based research: A decade of progress in education research? *Educational Researcher, 41*(1), 16–25.

Arhar, J., Niesz, T., Brossmann, J., Koebley, S., O'Brien, K., Loe, D., & Black, F. (2013). Creating a 'third space' in the context of a university–school partnership: Supporting teacher action research and the research preparation of doctoral students. *Educational Action Research, 21*(2), 218–236.

Barab, S., & Squire, K. (2004). Design-based research: Putting a stake in the ground. *The Journal of the Learning Sciences, 13*(1), 1–14.

Bevan, B., Penuel, W. R., Bell, P., & Buffington, P. J. (2018). Learning, generalizing, and local sense-making in research–practice partnerships. In B. Bevan & W. R. Penuel (Eds.), *Connecting research and practice for educational improvement* (pp. 17–30). Routledge.

Brydon-Miller, M., & Maguire, P. (2009). Participatory action research: Contributions to the development of practitioner inquiry in education. *Educational Action Research, 17*(1), 79–93.

Carlgren, I. (2012). The learning study as an approach for "clinical" subject matter didactic research. *International Journal for Lesson and Learning Studies, 1*(2), 126–139.

Carr, W., & Kemmis, S. (1986). *Becoming critical: Education knowledge and action research*. Routledge.

Carr, W., & Kemmis, S. (2005). Staying critical. *Educational action research, 13*(3), 347–358.

Cobb, P. A., Confrey, J., diSessa, A. A., Lehrer, R., & Schauble, L. (2003). Design experiments in educational research. *Educational Researcher, 32*(1), 9–13.

Coburn, C. E., & Penuel, W. R. (2016). Research–practice partnerships in education: Outcomes, dynamics, and open questions. *Educational Researcher, 45*(1), 48–54.

Coburn, C. E., Penuel, W. R., & Geil, K. E. (2013). *Research-practice partnerships: A strategy for leveraging research for educational improvement in school districts*. William T. Grant Foundation.

Cole, A. L. (1989). Researcher and teacher: Partners in theory building. *Journal of Education for Teaching, 75*(3), 225–237.

Cole, A. L., & Knowles, J. G. (1993). Teacher development partnership research: A focus on methods and issues. *American Educational Research Journal, 30*, 473–495.

Cornwall, A., & Jewkes, R. (1995). What is participatory research? *Social Science & Medicine, 41*(12), 1667–1676.

Dede, C. (2005). Commentary: The growing utilization of design-based research. *Contemporary Issues in Technology and Teacher Education, 5*(3), 345–348.

Dewey, J. (1929). *The sources of a science of education*. Liverright.

Farrell, C. C., Harrison, C., & Coburn, C. E. (2019). "What the hell is this, and who the hell are you?" Role and identity negotiation in research-practice partnerships. *AERA Open, 5*(2).

Hardy, I., Rönnerman, K., & Edwards-Groves, C. (2017). Transforming professional learning: Educational action research in practice. *European Educational Research Journal, 17*(3), 421–441.

Jull, J., Giles, A., & Graham, I. (2017). Community-based participatory research and integrated knowledge translation: Advancing the co-creation of knowledge. *Implementation Science, 12*(150).

Kemmis, S., & McTaggart, R. (2005). Participatory action research: Communicative action and the public sphere. In N. K. Denzin & Y. S. Lincoln (Eds.), *The Sage handbook of qualitative research* (pp. 559–603). Sage.

Kemmis, S., McTaggart, R., & Nixon, R. (2014). Introducing critical participatory action research. In S. Kemmis, R. McTaggart, & R. Nixon (Eds.), *The action research planner* (pp. 1–31). Springer.

Klein, E. J., Taylor, M., Onore, C., Strom, K., & Abrams, L. (2013). Finding a third space in teacher education: Creating an urban teacher residency. *Teaching Education, 24*(1), 27–57.
Lewin, K. (1946). Action research and minority problems. *Journal of Social Issues, 2*(4), 34–46.
Lillejord, S., & Børte, K. (2016). Partnership in teacher education – A research mapping. *European Journal of Teacher Education, 39*(5), 550–563.
Lundberg, B., & Starrin, B. (2001). *Participatory research. Tradition, theory and practice*. Karlstad University Press.
Magnusson, P., & Malmström, M. (2023). Practice-near school research in Sweden: Tendencies and teachers' roles. *Education Inquiry, 14*(3), 367–388.
Marton, F., & Runesson, U. (2014). The idea and practice of learning study. In *Realising Learning* (pp. 103–121). Routledge.
McIntyre, A. (2008). *Participatory action research.* Sage.
Nutley, S., Jung, T., & Walter, I. (2008). The many forms of research-informed practice: A framework for mapping diversity. *Cambridge Journal of Education, 38*(1), 53–71.
Reason, P., & Bradbury, H. (Eds.). (2001). *Handbook of action research: Participative inquiry and practice.*
Ryle, G. (1949). *The concept of mind.* Penguin Books Ltd..
Somekh, B. (2009). Agency through action research: Constructing active identities from theoretical models and metaphors. In S. Noffke & B. Somekh (Eds.), *The Sage handbook of educational action research* (pp. 370–380). Sage.
SOU 2018:19. *Utredningen om praktiknära skolforskning i samverkan [Inquiry on practice-near school research in collaboration*]. (2018). Forska tillsammans: samverkan för lärande och förbättring [Do research together: Collaboration for learning and improvement]. Norstedts juridik. https://www.riksdagen.se/sv/dokument-och-lagar/dokument/statens-offentliga-utredningar/forska-tillsammans-samverkan-for-larande-och_h6b319/html/
SOU 2024:81. *Ämneskunskaper och lärarskicklighet – en reformerad lärarutbildning.* [*Subject Knowledge and Teaching Skill: A Reformed Teacher Education Programme*] (2024). https://www.regeringen.se/contentassets/a1902cd0a0b94cee96972507f4a6c000/amneskunskaper-och-lararskicklighet%2D%2D-en-reformerad-lararutbildning-sou-202481.pdf
Stenhouse, L. (1975). *An introduction to curriculum research and development.* Heinemann.
Stetsenko, A. (2017). *The transformative mind.* In Expanding Vygotsky's approach to development and education. Cambridge University Press.
Williams, J. (2014). Teacher educator professional learning in the third space: Implications for identity and practice. *Journal of Teacher Education, 1–2.*
Zeichner, K. (2007). Accumulating knowledge across self-studies in teacher education. *Journal of Teacher Education, 58*, 36–46.
Zeichner, K. (2010). Rethinking the connections between campus courses and field experiences in College- and University-Based Teacher Education. *Journal of Teacher Education, 61*(1-2), 89–99.
Zeichner, K. M., & Noffke, S. E. (2001). Practitioner research. In V. Richardson (Ed.), *Handbook of research on teaching* (pp. 298–330). American Educational Research Association.

Mona Lundin is Professor of Education at the University of Gothenburg and an internationally active researcher in the field of digitalization, education, and professional learning. Her work draws on sociocultural perspectives to explore how digital technologies shape teaching practices, learning environments, and professional development—particularly within teacher education. She has played a leading role in numerous externally funded research projects, including as co-applicant in the current Swedish Research Council project "The search for professional digital competence in Swedish teacher education" (2025–2028), and in Teachers' professional digital competence (2020–2023). Lundin has also led research on flipped-classroom practices and participated in Nordic collaborative projects on digital welfare, childhood digitalization and higher education research. Her research has been widely published, and she contributes actively to national and international networks.

Annika Lantz-Andersson is Professor in Education at the University of Gothenburg. Her research is grounded in sociohistorical and sociomaterial traditions that emphasise how humans co-create and reshape socio-technical contexts. She investigates the educational consequences of digitalization, with particular attention to how teaching, learning and the teaching profession are reshaped when human expertise interacts with digital technologies. Across multiple educational stages, she has studied technology-supported teaching. Methodologically, her repertoire includes design-based and practice-based educational research, alongside thematic, interactional, and interview analyses. Current projects examine teachers' professional digital competence and how prevailing approaches to participation, learning, and competence might be extended.

Ewa Skantz holds a PhD in Education and is a senior lecturer at the Department of Education, Communication and Learning, University of Gothenburg. An overall research interest is how digitalization changes the conditions for social interaction and communication, formal teaching, and learning. Her research particularly concerns practice-based research and teachers' professional digital competence. A further topic is ocean literacy and marine education.

Pia Williams is a Professor of Child and Youth Studies at the University of Gothenburg. Her research focuses on early childhood education and care, with a particular interest in the conditions that promote young children's learning and development in preschool settings. She leads and collaborates on research projects exploring educational quality, aiming to evaluate, develop, and ensure high-quality preschool education. A key focus of her work is bridging research and practice to support the development of pedagogical strategies that enhance outcomes for children in early educational contexts.

Advancing Simulation Pedagogies in MET

Charlott Sellberg

Abstract This final chapter offers a reflective synthesis of the preceding contributions, drawing together insights from research and practice to consider how simulation pedagogies in maritime education and training (MET) are currently enacted, and how they might be further developed. Against the backdrop of an industry marked by rapid technological innovation, increasing internationalisation, and intensified demands for professional competence, the chapter examines how simulation-based training is situated within broader educational and institutional frameworks. The discussion foregrounds how pedagogical practices are shaped by local conditions, disciplinary traditions, and industry standards. By revisiting key themes across the book, this chapter argues for a pedagogically grounded approach to simulation. Finally, the chapter outlines several directions for developing simulation pedagogies in MET, pointing toward a future in which simulation is not only a site for training, but a space for engaging with the complexities of professional learning in an increasingly technically advanced and globalised maritime sector.

Keywords Simulation-based training · Pedagogical development · Maritime autonomous surface ships (MASS) · Artificial intelligence (AI) · Maritime education and training (MET)

1 Towards Shared Repertoires of Simulation Pedagogies

Simulation has become a cornerstone of maritime education and training (MET), providing structured and safe environments for training complex and high-risk work practices (Giskeødegård & Wahl, 2025). Across the chapters of this book,

C. Sellberg (✉)
Department of Applied Information Technology, University of Gothenburg,
Gothenburg, Sweden
e-mail: charlott.sellberg@ait.gu.se

C. Sellberg (ed.), *Simulations in Maritime Education and Training*, WMU Studies in Maritime Affairs 12, https://doi.org/10.1007/978-3-032-08394-4_8

simulation pedagogies in MET are approached as situated practices of teaching and learning. Through these practices, instructors, students, artefacts, and environments are brought together in ways that shape how professional knowledge is enacted, negotiated, and developed (Ekse et al., 2025). This orchestration of learning involves aligning learning objectives with simulator scenarios, framing the activity in pedagogically meaningful ways, working with artefacts such as charts, checklists, and instruments to support learning, and attending to how the simulation environment affords, or constrains, particular forms of participation. In this view, simulation-based training emerges as a space for the coordination of multiple forms of knowing in action, a space where professional judgement, communication, and situational awareness are developed through embodied interaction within a socio-technical system (Sellberg & Lundin, 2018).

The first chapter situates simulation within the historical development of maritime education, tracing its evolution from apprenticeship-based models toward increasingly formalised and technologically mediated forms of training. Simulators, initially introduced for technical skill acquisition, have become integrated into the very signature pedagogies of MET. What is noteworthy is the layered conceptualisation of simulator fidelity, encompassing both technical, psychological, and interactional fidelity. Instructors are invited to reflect not only on what the simulator looks like, but also on the kinds of actions, relations, and learning it enables. This chapter also highlights the scarcity of theoretically informed pedagogical perspectives in the MET literature, where the focus tends to remain on human factors or instrumentally framed learning outcomes. A key contribution is the call for an advanced understanding of the relationship between learning goals, scenario design, and simulator modality, an understanding that moves beyond fidelity as realism and toward fidelity as *pedagogical relevance*. The chapter ultimately calls attention to the importance of pedagogical reasoning in simulator design and use, pushing back against the tendency to treat simulators as self-evident tools for skill transfer.

Building on the recognition that effective simulation is more than access to advanced equipment, chapter "Designing Simulation for Learning" focuses on scenario design as a pedagogical practice. The main argument is that instructors' professional experience, while invaluable, needs to be supported by systematic instructional design models to foster professional learning. Using the Four-Component Instructional Design model (4C/ID), the chapter offers a practical and theoretically grounded framework for connecting learning objectives, scenario content, and student development. What makes this contribution particularly relevant is its attentiveness to the challenges instructors face in practice, how to tailor the complexity of scenarios to students' levels, how to define clear and assessable learning objectives, and how to avoid fragmenting learning into disconnected tasks. The chapter thus bridges experiential and theory-informed approaches, showing how design practices in MET can support not only competence development, but also students' ability to reflect, make decisions, and act with professional judgment in unforeseen situations. In doing so, the chapter frames simulation as a resource for developing professional agency.

Chapter "Preparing for Simulation Through Briefing" turns attention to the preparatory practices that precede the simulation scenario itself, positioning the briefing phase as a critical, yet often overlooked, component of simulation pedagogy. Drawing on empirical examples and pedagogical theory, the authors show how briefings are not just informational sessions, but spaces where expectations are negotiated, ground rules established, and tasks made meaningful. The notion of the fiction contract is introduced to highlight how students must be supported in treating simulations as both real and artificial, as professional tasks with pedagogical intent. The briefing is thus framed as a site of sense-making, not merely about what students are to do, but how they are to understand their roles, the task, and its relevance to professional practice. Instructors are invited to move from demonstration as "showing" to demonstration as "involving", supporting students' active orientation to the learning activity. In this way, the chapter expands the conception of briefing beyond content delivery, highlighting its role in shaping how students approach the simulation as a site for learning and development.

Chapter "The Instructor's Work During Simulations" brings focus to the often-invisible work of instructors during the simulation scenario itself, a phase that sits between the activities of pre-simulation briefing and post-simulation debriefing. Harrington outlines how instructors manage the complex demands of monitoring, intervening, and supporting student learning in real time, drawing on professional vision and pedagogical judgment to interpret unfolding events and scaffold performance. In this view, simulation here is a space of emergent activity, requiring instructors to respond flexibly as students engage with the scenario in unexpected ways. Importantly, the chapter conceptualises the instructor's role as both instructional and developmental. Drawing on the framework of threshold concepts, simulation is situated as a messy, nonlinear process through which learners begin to take on the epistemic commitments of the profession. The instructor's role, then, is to support students in their ongoing process of becoming through navigating liminal spaces where professional identities are still in formation. The chapter offers a conceptual language for articulating how pedagogical decisions in simulation are entangled with broader educational goals, including the cultivation of professional judgment, role-taking, and collaborative competencies. Harrington also describes three mechanisms through which instructors contribute to professional learning during simulation: (1) filling in missing aspects of the work context, (2) sustaining an "as-if" orientation to professional roles, and (3) making visible what counts as good seamanship in particular situations. These mechanisms point to simulation as a site where professional norms are not only practised but interpreted and negotiated, often in subtle and embodied ways. In doing so, the chapter offers a powerful reminder that simulation pedagogies are shaped as much by the micro-interactions of instructional practice as by formal standards or curriculum documents.

Chapter "Facilitating debriefings for feedback, discussion, and reflection" expands the pedagogical lens further, shifting attention to the debriefing phase, sometimes described as the most critical moment for learning in simulation-based training. Drawing on a range of models from healthcare and high-risk domains, the chapter conceptualises debriefing as a structured, dialogic space where learners can reconstruct experience, examine assumptions, and engage in collective sense-making. Importantly, the chapter underscores that debriefing is not reducible to instructor-led feedback. Rather, it is a negotiated practice that depends on psychological safety, shared reflection, and a clear alignment with learning objectives. Multiple debriefing frameworks are introduced, including the Reaction–Analysis–Summary model, the 3D model (Defusing–Discovering–Deepening), and the Diamond Debrief, each offering different orientations to timing, emotional expression, and conceptual integration. These models are not presented as prescriptions, but as pedagogical tools that instructors can adapt in relation to their learners, scenarios, and institutional constraints. What runs across all of them is the view of debriefing as a structured yet open-ended pedagogical activity, in which meaning is made not only of the task completed, but of what it reveals about professional reasoning, coordination, and decision-making. Chapter "Facilitating debriefings for feedback, discussion, and reflection" also draws attention to the practical and institutional conditions that shape debriefing, including time constraints, access to mediational tools (e.g., video, playback, maps), and the instructor's own level of pedagogical training. In this way, debriefing is framed as a professional practice that relies on the instructor's ability to move between evaluative and instructional stances, to balance emotional and cognitive responses, and to connect individual performance to general norms of what constitutes professional competence in the maritime domain. Sharma's chapter thus brings into focus the reflective dimensions of simulation pedagogies and the critical role of debriefing in supporting learning transfer, generalisation, and professional identity formation.

Chapter "Assessing Competence Through Simulation" turns the focus to assessment, a domain where the pedagogical and regulatory logics of MET most clearly intersect. In this chapter, Sellberg examines how simulators are utilized not only for teaching but also for high-stakes evaluation of seafarer competence, and the tensions this creates regarding validity, reliability, and fairness. While the STCW Convention sets the formal requirements for certification, the chapter documents how instructors grapple with implementing these requirements in practice, particularly in scenarios that involve tacit knowledge, embodied skills, and collaborative problem-solving. A key theme is the tension between the push for standardisation and the inherently situated, interpretive nature of professional performance. Assessment in simulation cannot be reduced to binary judgments or decontextualised checklists; it requires instructors to draw on their own professional vision, to interpret actions in context, and to decide what counts as competent conduct. Drawing on empirical studies of assessment situations, the chapter shows how instructors often intervene during tests to clarify questions, rephrase tasks, or guide students through breakdowns, not to compromise the test, but to enable students to show what they know. In this sense, assessment becomes a formative as well as summative activity, embedded in a broader process of professional socialisation. Moreover, the chapter highlights the risk of bias in assessment, from gut feelings and halo effects to more structural inequities that arise when students' confidence or communication styles are mistaken for competence. Rather than proposing easy fixes, the chapter invites a reflective and collective approach to assessment practice, including assessor training, collaborative calibration, and critical discussion of how professional standards are enacted in specific situations. Assessment, in this view, is both a pedagogical and ethical activity, a space where values about what it means to be a seafarer are made visible and consequential.

2 Advancing Simulator Pedagogy in MET

Chapter "Instructors as Co-Researchers: Strengthening Professional Practice Through Collaborative Research" shifts attention from the simulation room to the broader conditions under which instructors engage in professional development and pedagogical inquiry. Drawing on traditions such as Action Research, Participatory Action Research, and Design-Based Research, Lundin et al. explore how maritime instructors can take an active role as co-researchers in collaborative educational research. The chapter foregrounds the value of practice-centred inquiry as a form of knowledge production grounded in the questions, dilemmas, and routines of instructional work. A key contribution is the conceptual framing of research-practice partnerships as work in a *third space*, a hybrid arena where academic and professional knowledge are not only exchanged but jointly developed. In this space, instructors and researchers are positioned not as knowledge givers and receivers, but as colleagues who bring different forms of expertise to bear on shared problems of practice. The chapter is careful, however, not to idealise this collaboration. It documents the institutional, epistemic, and interpersonal tensions that often arise, including role ambiguity, power asymmetries, and conflicting timelines, and argues for the importance of sustained, reciprocal relationships built on trust, dialogue, and shared ownership.

Particularly relevant for MET is the chapter's attention to the structural challenges of professional development in post-secondary vocational education. Instructors often have strong subject matter expertise but limited formal training in pedagogy or research. Yet, as the chapter shows, when given the opportunity to investigate their own teaching, through video analysis, lesson study, or joint curriculum development, instructors not only refine their own practice, but also contribute to the broader knowledge base of teaching. Collaborative research in this sense becomes a professionalising force: a way of developing pedagogical language, expanding instructional repertoires, and making the complexities of simulation-based education visible and discussable. In linking research and teaching in this way, the chapter also challenges us to rethink how pedagogical development is organised and supported in MET. Rather than relying solely on one-off courses or externally driven projects, Lundin et al. argue for embedding inquiry into the everyday work of instructors, for creating institutional arrangements where pedagogical development is an integral part of what it means to be a professional educator. In doing so, the chapter points to a broader vision of simulation pedagogy as an evolving, research-informed practice sustained by collective reflection and shared responsibility.

At the same time, simulation in MET is closely tied to regulatory frameworks such as the *International Convention on Standards of Training, Certification and Watchkeeping for Seafarers* (STCW), which prescribe learning objectives and specify simulator types, thereby shaping what is possible and expected in educational practice. While the STCW standards provide an important structure, they also risk narrowing the pedagogical focus to compliance and certification, favouring measurable performance over reflective practice, standardisation over contextual responsiveness, and procedural correctness over the cultivation of professional judgment (Sellberg et al., 2022). What the chapters in this book collectively point to is the need to balance these regulatory demands with the tacit, situated, and socially organised dimensions of maritime knowledge and expertise.

Moreover, it is essential to acknowledge that MET institutions must also navigate and align with broader higher education policies and standards, which impact curriculum design, assessment approaches, academic quality assurance, and institutional accreditation (Manuel, 2017). This dual regulatory landscape requires MET programs to balance industry-specific training requirements with the policies established by higher education authorities. On one hand, aligning with STCW ensures that graduates meet internationally recognized competency standards essential for safe and effective maritime operations. On the other hand, the imperatives of higher education emphasize critical thinking, academic rigor, and the development of transferable skills that extend beyond immediate industry needs (Kaiser et al., 2014). Consequently, MET programs must reconcile these potentially competing priorities by fostering curricula and pedagogical approaches that satisfy industry requirements while also promoting critical thinking, reflective learning, and professional identity formation. In practice, this means that maritime educators are encouraged to create authentic learning environments that enable students to engage with uncertainty, interpret complex situations, and negotiate meaning with peers and instructors (Ghosh & Emad, 2024). This approach aligns with contemporary higher education goals of cultivating lifelong learning and adaptability, qualities increasingly critical in a maritime industry characterized by rapid technological change and evolving operational contexts.

3 Towards Increased Levels of Digitalisation and Automation

The ongoing development and implementation of Maritime Autonomous Surface Ships (MASS) marks a profound shift in the nature of maritime work (Belabyad et al., 2025). As autonomous navigation systems, AI-driven decision support, and integrated sensor networks become more prevalent on board and ashore, the role of the seafarer is being redefined. For example, with reduced or remote manning, seafarers may no longer be physically present on board but operate from shore-based control centres. This spatial and organisational separation raises questions about how tacit dimensions of knowledge, collaborative forms of learning, and professional judgment enter into these new forms of highly automated work environments (Sellberg et al., 2022). Rather than performing manual operations, seafarers are increasingly expected to monitor, manage, and intervene in automated systems, especially in cases where these systems reach their operational limits or encounter novel, ambiguous situations (Endsley, 2017). This transition involves a deeper transformation of professional competence, and the emerging literature on MASS consistently highlights the need for new skillsets that extend beyond traditional technical and operational training (Ghosh & Emad, 2024). These include complex problem-solving, systems thinking, digital and data literacy, human-machine interaction, ethical reasoning, and adaptive decision-making under uncertainty (Table 1).

As human roles evolve to include the oversight of autonomous systems, new demands emerge, for example, towards understanding how these systems function, knowing when to intervene or trust automation, and communicating effectively within hybrid teams of humans and machines (O'Neill et al., 2022). Moreover, the move towards increased automation introduces new challenges in terms of responsibility allocation and regulatory compliance (Palbar Misas et al., 2024). In this context, so-called twenty-first-century competencies are essential components of safe and effective maritime practice in an age of digitalisation (Pazaver & Kitada, 2025). Overall, the MASS literature underscores the need for MET institutions to take a proactive role in bridging the gap between technological innovation and pedagogical adaptation, ensuring that future maritime professionals are not only technically skilled but also critically reflective and ethically grounded actors in increasingly automated workspaces.

At the same time, MET is also undergoing a process of digitalisation, with new educational technologies, including AI-powered chatbots and learning analytics, increasingly entering pedagogical practices (Karimi et al., 2024). Intelligent tutoring systems, adaptive feedback mechanisms, and data-driven performance monitoring are being introduced to personalise learning trajectories, identify areas for improvement, and support instructors in making informed pedagogical decisions (Odéen et al., 2024). These developments offer opportunities to enhance the effectiveness and responsiveness of simulation-based training, for instance, by analysing patterns in student behaviour, decision-making, and communication during scenarios through a variety of sensors (Sellberg & Sharma, 2025). However, the integration of such technologies also raises important questions about pedagogical intent, data ethics, and the role of human judgment in instruction, feedback, and assessment. As MET institutions explore these new possibilities, it is essential to ensure that educational practices remain grounded in professionally meaningful learning objectives and that technological tools are used to complement, rather than replace, the relational, interpretive, and context-sensitive dimensions of professional learning in simulation-based training. Thus, simulation pedagogy in MET needs to enable learners to critically engage with and reflect on their experiences under increasingly complex technological conditions. This reflective engagement requires not only technical proficiency but also new literacies necessary for navigating both current and future demands.

Table 1 Skills and Competencies for Operating MASS (adapted from Belabyad et al., 2025)

Competency Category	Specific Competency Area	Skills and Competencies
Technical & Digital	Digital Competencies	Cybersecurity, Data Analytics, Cloud Computing
Technical & Digital	Automations & Systems	Automation Engineering, System Monitoring, Troubleshooting
Technical & Digital	Engineering Competencies	Electrical/Electronic/Mechanical/Software/Systems/Sensor Engineering
Operational	Remote Maritime Operations	Monitoring, Multi-tasking, Operational Limitations Awareness
Operational	Maritime Expertise	COLREGs, Radar, ECDIS, Legal and Regulatory Knowledge
Operational	Safety Management	Safety Protocols, Situational Awareness, Risk Assessment
Higher-Order Thinking	Critical Thinking	Analytical Reasoning, Data Evaluation, Root Cause Analysis
Higher-Order Thinking	Decision-Making & Problem Solving	Judgment under Stress, Data-Driven Decisions, AI Knowledge
Interpersonal	Communication & Collaboration	Virtual Teaming, Verbal/Multilingual Communication, Emotional Intelligence
Interpersonal	Adaptability & Flexibility	Tech Adaptation, Cultural Awareness, Resilience, Lifelong Learning
Interpersonal	Leadership	Leadership, Team Management, Stakeholder Coordination

4 Discussion

This chapter revisits the book's contributions to outline how simulation pedagogies in maritime education can be further developed. Across the chapters, simulation is approached as a pedagogical practice shaped by professional knowledge, regulatory frameworks, and local conditions. Taken together, the contributions point to simulation as a site where professional knowledge is enacted, negotiated, and made learnable through collaborative work between instructors and students. Supporting this work requires attention to pedagogical reasoning, that is, a focus on how simulation scenarios are designed, framed, scaffolded, and assessed in ways that align with broader educational aims. In doing so, the book also opens up questions about the future of maritime education in light of the changing conditions of the maritime industry. As technologies evolve, as operations become increasingly automated and distributed, and as the expectations placed on future seafarers shift, the role of simulation in preparing students for maritime work will need to be revisited. In this sense, the book not only documents current practice but contributes to an emerging pedagogical discourse that recognises simulation as a central, and evolving, component in the professional formation of maritime professionals.

At the same time, the book leaves important areas underexplored. In particular, the relationship between simulation-based and on-board training is not addressed, even though this remains a key concern in MET programmes. As Mori and Manuel (2024) have shown, on-board training arrangements vary significantly across regions and institutions, shaped by differing rationales and practical constraints. These arrangements fall into at least seven dominant types, based on how simulator training substitutes or complements sea time, how on-board training is integrated into the academic programme, the scheduling of placements, and the kinds of ships used. Too often, simulator time and sea time are framed as interchangeable, where one is expected to compensate for a lack of the other. However, I argue that this view risks overlooking the distinctive pedagogical affordances of each setting. While simulators offer controlled, repeatable scenarios ideal for focused practice, instructor feedback, and experimentation, on-board training can provide access to the everyday complexity, unpredictability, and social norms of the profession *if* effectively structured (see Mori, 2025, for an extended discussion). These forms of training support different kinds of learning and should be seen as complementary rather than equivalent. One does not replace the other; rather, each setting provides meaning to the other. While simulations help students make sense of experiences at sea, time on board supports students to acknowledge the relevance (or irrelevance) of aspects of simulated scenarios. Hence, understanding how these settings should be pedagogically integrated is an area that demands further attention in MET research.

While this book provides a comprehensive account of simulation-based training from the perspective of instructors, it does not directly engage with students' experiences or perceptions of simulation-based training. Yet studies that explore how students perceive simulations offer valuable insights into why pedagogical approaches may need to change. For example, in Harrington et al. (2025), workshop discussions reveal that students often do not recognize reflective debriefing discussions as feedback, even in simulated learning environments where reflection is central. Instead, the students tended to seek concrete answers about what is correct and what is not, expressing a desire for explicit and personalized instructions on how to improve. These findings highlight a potential misalignment between current teaching practices and students' expectations, suggesting that effective simulation pedagogy needs to account for how students understand and respond to feedback in order to support meaningful learning, for example, through meta-discussions on why we debrief and what can be learned from the mistakes of peers (Lymer et al., 2025).

It is also important to acknowledge that while this book has explored the different phases of simulation-based training, from briefing and scenario execution to debriefing, assessment, and pedagogical development, the book has not scrutinized the specific learning objectives and pedagogical considerations tied to particular courses or certifications. For instance, we have not provided in-depth discussions of how simulation is integrated into basic safety training, basic navigation courses, bridge resource management (BRM), dynamic positioning (DP), or other specialized modules that form critical components of MET. Each of these areas involves distinct competencies, regulatory requirements, and pedagogical challenges that shape how simulations are designed, enacted, and assessed. By not engaging with these course-specific dynamics in detail, this volume leaves open important questions about how general insights into simulation pedagogy translate into the complex realities of particular training contexts. Future handbooks could usefully take up this task, offering more granular analyses of how simulation-based training supports the development of technical, cognitive, and collaborative skills across different areas of maritime education.

Another limitation of the book is that its focus is largely limited to Sweden and Norway, and although the insights are of broader relevance, they reflect particular institutional and regulatory conditions. The geographic scope means that key questions about how simulation pedagogies unfold in other maritime education contexts, shaped by different educational practices, infrastructures, and policy environments, remain largely unanswered in this book. While some degree of uniformity seems to exist, largely because institutions share a common understanding of the STCW Convention and IMO model courses, differences emerge in the implementation due to varying interpretations, available infrastructure, and institutional priorities (Nazir et al., 2019). For example, Nazir et al. (2019) found that feedback and assessment practices differ across settings, mostly based on the instructors' expertise and available technological support. These findings suggest that simulation pedagogies are deeply influenced by contextual factors beyond formal guidelines, resulting in a heterogeneous landscape across maritime training institutions. While this diversity enables local adaptation and pedagogical innovation, it also raises questions about consistency, quality assurance, and the broader goal of international harmonization in maritime education that needs further investigation. In particular, as the maritime industry is inherently a global one, there is a need for continuous dialogue across national and institutional boundaries. This includes working toward international research collaborations and institutional partnerships that can support the mutual development of simulation pedagogies in a rapidly changing world.

One promising initiative in this regard is the International Association of Maritime Universities (IAMU) Simulator Group, which brings together educators and MET researchers from around the world to share experiences, develop common frameworks, and explore pedagogical innovations in simulation-based training. As a forum grounded in both research and practice, the IAMU Simulator Group has the potential to support comparative studies, coordinate joint projects, and build a shared language for discussing simulation as a pedagogical practice. By providing a platform for sustained collaboration, the group exemplifies the kind of global engagement needed to move beyond regionally bounded perspectives of simulation pedagogy in maritime education.

References

Belabyad, M., Kontovas, C., Pyne, R., & Chang, C.-H. (2025). Skills and competencies for operating maritime autonomous surface ships (MASS): A systematic review and bibliometric analysis. *Maritime Policy & Management*, 1–26.

Belabyad, M., Kontovas, C., Pyne, R., Shi, W., Li, N., Szwed, P., & Chang, C. H. (2025). The human element in autonomous shipping: A study on skills and competency requirements. *WMU Journal of Maritime Affairs*, 1–31.

Ekse, E. B., Giskeødegård, M. F., Sellberg, C., & Wahl, A. M. (2025). Cultivating students' professional skills through simulation-based training: Nautical, nurse, and biomedical laboratory science teachers' perspectives. *Discover Education, 4*, 225.

Endsley, M. R. (2017). From here to autonomy: Lessons learned from human–automation research. *Human Factors, 59*(1), 5–27.

Ghosh, S., & Emad, G. R. (2024). Identifying challenges in designing and implementing a skills and competency framework for future seafarers: A systematic literature review. *Australian Journal of Maritime & Ocean Affairs*, 1–14.

Giskeødegård, M. F., & Wahl, A. M. (2025). Signature pedagogies of Norwegian simulator-based nautical training. *Discover Education, 4*(1), 148.

Harrington, S., Klein, S., Fratoni, A. C., Sellberg, C., & Lindwall, O. (2025). Meaningful automated feedback? Balancing tensions in participatory learning analytics design. *Proceedings from the 18th international conference on computer-supported collaborative learning, Helsinki, Finland.*

Kaiser, F., Maassen, P., Meek, L., van Vught, F., de Weert, E., & Goedegebuure, L. (Eds.). (2014). *Higher education policy: An international comparative perspective*. Elsevier.

Karimi, E., Smith, J., Billard, R., & Veitch, B. (2024). AI-based adaptive instructional systems for maritime safety training: A systematic literature review. *Discover Artificial Intelligence, 4*(1), 51.

Lymer, G., Lindwall, O., & Sellberg, C. (2025). Laughter, failure talk, and the sensitive nature of negative feedback. *Classroom Discourse*, 1–26.

Manuel, M. E. (2017). Vocational and academic approaches to maritime education and training (MET): Trends, challenges and opportunities. *WMU Journal of Maritime Affairs, 16*(3), 473–483.

Mori, Y. (2025). An analysis of challenges and impacts of different trends/factors on the implementation of onboard training: Perspectives from maritime education and training institutions. *Cognition, Technology & Work, 27*, 1–15.

Mori, Y., & Manuel, M. (2024). An exploratory study of approaches to onboard training in different countries. *Journal of International Maritime Safety, Environmental Affairs, and Shipping, 8*(3), 2386221.

Nazir, S., Jungefeldt, S., & Sharma, A. (2019). Maritime simulator training across Europe: A comparative study. *WMU Journal of Maritime Affairs, 18*(1), 197–224.

O'neill, T., McNeese, N., Barron, A., & Schelble, B. (2022). Human–autonomy teaming: A review and analysis of the empirical literature. *Human Factors, 64*(5), 904–938.

Odéen, M., Sellberg, C., Praetorius, G., Lindwall, O., Englund, L., & Andersson, A. (2024). Intelligent learning Systems for Simulator-Based Professional Training: A systematic literature review. In *International conference in methodologies and intelligent Systems for Technology Enhanced Learning* (pp. 109–119). Cham.

Palbar Misas, J. D., Hopcraft, R., Tam, K., & Jones, K. (2024). Future of maritime autonomy: Cybersecurity, trust and mariner's situational awareness. *Journal of Marine Engineering & Technology, 23*(3), 224–235.

Pazaver, A., & Kitada, M. (2025). Integrating twenty-first century skills into STCW competences: Implications for maritime education and training. *WMU Journal of Maritime Affairs, 24*, 1–22.

Sellberg, C., & Lundin, M. (2018). Tasks and instructions on the simulated bridge: Discourses of temporality in maritime training. *Discourse Studies, 20*(2), 289–305.

Sellberg, C., & Sharma, A. (2025). Toward multimodal learning analytics in simulation-based collaborative learning: A design ethnography of maritime training. *International Journal of Computer-Supported Collaborative Learning, 20*(2), 201–221.

Sellberg, C., Wahl, A. M., & Viktorelius, M. (2022). Moving control from ship to shore: Computer supported cooperative work in autonomous shipping. In *Future of information and communication conference* (pp. 123–134). Springer International Publishing.

Charlott Sellberg is an Associate Professor in IT and Learning at the Department of Applied IT, University of Gothenburg. With a multidisciplinary background in Cognitive Science and Human-Computer Interaction, she earned a PhD in Education from the University of Gothenburg in 2018. Her doctoral work examined both the theoretical and practical aspects of simulation-based training in the maritime industry. Since then, she has continued to study the role of simulations in maritime training and assessment, in areas such as basic safety training, SAR operations, and advanced ship handling, continuing to deepen the understanding of simulation as a learning environment and informing educational practices within the maritime domain.

MIX
Papier aus verantwortungsvollen Quellen
Paper from responsible sources
FSC® C105338

If you have any concerns about our products,
you can contact us on
ProductSafety@springernature.com

In case Publisher is established outside the EU,
the EU authorized representative is:
Springer Nature Customer Service Center GmbH
Europaplatz 3, 69115 Heidelberg, Germany

Printed by Libri Plureos GmbH
in Hamburg, Germany